INTRODUCTION TO GEOCHEMISTRY

GEOPHYSICS AND
ASTROPHYSICS MONOGRAPHS

AN INTERNATIONAL SERIES OF FUNDAMENTAL TEXTBOOKS

VOLUME 10

INTRODUCTION
TO GEOCHEMISTRY

by

CLAUDE-JEAN ALLÈGRE

Department of Earth Sciences, University of Paris 7

and

GIL MICHARD

Department of Chemistry, University of Paris 7

D. REIDEL PUBLISHING COMPANY

DORDRECHT-HOLLAND / BOSTON-U.S.A.

INTRODUCTION À LA GÉOCHIMIE

First published by Presses Universitaires de France, Paris, 1973

Translated from the French by Robert N. Varney

Library of Congress Catalog Card Number 74–83871

Cloth edition: ISBN 90 277 0497 X
Paperback edition: ISBN 90 277 0498 8

Published by D. Reidel Publishing Company,
P.O. Box 17, Dordrecht, Holland

Sold and distributed in the U.S.A., Canada, and Mexico
by D. Reidel Publishing Company, Inc.
306 Dartmouth Street, Boston,
Mass. 02116, U.S.A.

TABLE OF CONTENTS

PREFACE

The study of chemistry has, for many years, included the study of natural minerals; the term *geochemistry* was introduced by Schönbein in 1838, and the laboratory synthesis of geological objects also began in the nineteenth century (Daubrée). Thereafter, the paths of chemists and geologists diverged.

The geochemistry of the twentieth century began with the establishing of the mean composition of the Earth's crust through analysis of a large number of rock specimens. The first tabulation is due to Clarke in 1908, and it is continuously being amended to this day.

Next, an attempt was made to establish the balance of interchanges of elements among the different parts of the Earth, and Vernadski introduced the concept of the geochemical cycle.

Then physical chemistry appeared on the scene of Earth science. The methods of thermodynamics were applied to the study of the formation of minerals. The ideas concerning the structure of atoms and their combinations into molecules and crystals opened new avenues to mineralogy. The discovery of stable and radioactive isotopes cleared the way for the quantitative determination of the temperature of formation and the ages of the rocks. The inaccessible internal structures, beyond reach of direct observation, became observable through the methods of geophysics.

Today, there are numerous approaches to geochemistry and publications now fill dozens of specialized journals. It is in fact not possible to give an overall view of geochemistry, however superficially. Before we can start at all, we shall show how the methods of chemistry and of chemical physics can be expanded to deal with the great laboratory that is the whole Earth. Some of the methods have already demonstrated their effectiveness in Earth science. Other, newer techniques hold great promise but may themselves be rapidly surpassed.

The features of the Earth are the messages that have been written a long time ago, perhaps at inaccessible depths within the globe. To decipher their meanings, considerable imagination may be required – and validated –, and it is desirable that the methods of physical chemistry should guide and also control the imagination.

In a work of this sort, rules concerning references are essential. Here are our rules:

(1) A bibliography appears at the end of each chapter.

(2) The references cited, however debatable our choice may seem to be, are necessitated by the plan of the book.

(3) Authors cited in the text are not necessarily listed in the bibliography. This sort of omission occurs whenever we choose to identify the origin of a discovery but prefer a more readable review than the original reference for the bibliography.

PREFACE BY THE TRANSLATOR

The rendering of the French text by Drs Allègre and Michard into English has been done without amendments, editings, or alterations except for the correction of a few misprints. American rather than British usages have been chosen; for example, one billion is 10^9. The French authors have been furnished with the translation and have corrected errors in terminology and have added several paragraphs and figures that will thus be new in the English edition. The translator has been assisted by Professor W. R. Evitt of the Department of Geology of Stanford University who kindly read every word of both the French and the English texts and made many improvements in the latter. The translator expresses his unbounded gratitude to Professor Evitt. The translator is also deeply indebted to Dr B. M. McCormac for much necessary work in editing the English manuscript.

ROBERT N. VARNEY

Palo Alto, California
July 31, 1974

CHAPTER 1

THE EARTH AS A CHEMICAL SYSTEM

1.1. The Earth within the Solar System

The solar system is composed of the Sun, which provides 99.8% of the total mass, the planets, and the 'planetary objects' (dust, comets, asteroids).

The planets travel around the Sun in elliptical orbits. A distinction is made between the minor planets close to the Sun – Mercury, Venus, Earth, Mars – and the major planets – Jupiter, Saturn, Uranus, Neptune, and Pluto. Between the two groups, billions of asteroids circulate in their orbits around the Sun (the asteroid belt). They vary greatly in size; the largest, Ceres, has a diameter of 770 km. It is estimated that there are 10^4 asteroids larger than 10 km in diameter and 10^{11} larger than 1 m. Some of the orbits cut across planetary orbits. Among the meteorites that have struck the Earth, two were found to have had orbits that were within the scope of the asteroid belt. This number may seem small, but the number of successful identifications of meteorite orbits prior to impact is also very small.

The Earth is thus a member of the solar system, and one can compare samples of it with meteorites and with samples which may be collected from other planets or satellites. These comparisons, begun by Urey (1952) have great importance in recent developments of geochemistry and cosmological chemistry.

Only the surfaces of planets are directly accessible to us, either for recovery of specimens or for spectral analysis. The interiors of the planets will only become known to us indirectly, either by the construction of models that explain the formation of the surface features or by the comparison with meteorites which are the only cosmic objects that can be studied in their entirety.

1.2. The Scale of Time

The solar system evolved in bygone times that are measurable in billions (10^9) of years.

It is believed that the creation of the chemical elements of the solar system began 7 ot 8 billion years ago.

The formation of the planets would have occurred in a relatively short time interval, 4.6 billion years ago. Life first appeared on Earth around 3.5 billion years ago, and man about 0.002 billion years ago.

It should also be noted that the geological eras (primary, secondary, tertiary, quaternary), on which all of *classical* geology is built, cover only the last five hundred million years (Figure 1-1).

Nucleosynthesis
|
Formation of the
Earth, planets,
and meteorites   Start of   Appearance of
/ Start of life   fossiliferous age   man

7   4.6   1 billion   100 millions   10 millions   1million
billions

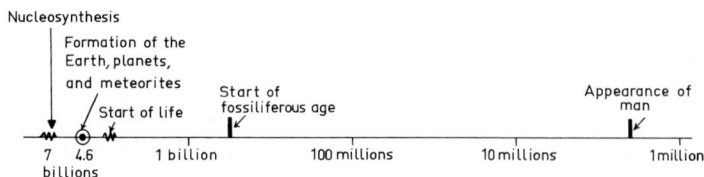

Fig. 1.1.   Time scale for planetary events.

1.3. The Overall Chemical Composition of the Earth, the Solar System, and the Universe

Spectroscopic examination of the Sun and the nearer stars (made easier in recent times by space vehicles) as well as chemical analyses of meteorites (those rocks that fall out of the sky!) have permitted calculation of the abundances of the various chemical elements in the solar system and nearby space. Depending on the choice of stars and the choice of statistical methods employed, the absolute abundances vary from one author to another. Nevertheless, when one plots the abundances on a logarithmic scale (relative to a value for silicon arbitrarily set at 10000) against atomic number of the elements, some absolutely general regularities emerge (Figure 1-2).

(1) As the atomic number (Z) increases, the abundance decreases very rapidly.

(2) For elements with atomic number greater than 45, the abundances are virtually constant.

(3) The elements with even atomic number are much more abundant than those with odd atomic number. This is known as the law of Oddo and Harkins.

(4) Three elements with low atomic number have a disproportionately low abundance considering their Z values; they are Li, Be, and B.

(5) Iron by contrast shows excessive abundance for its atomic number.

The explanation of all these phenomena is to be sought in nuclear astrophysics and in the phenomena of nuclear synthesis. This subject falls outside of the scope of this work, but we shall give several illustrations that will enable the reader to grasp the meaning of our statements.

The nucleus is composed as a first approximation of neutrons and protons. Two nuclei having the same atomic number but different mass numbers are called isotopes. Naturally occurring elements are mixtures of isotopes. Let us consider the rare earths; the curve of their abundance is a zigzag that can readily be explained by the fact that odd numbered atoms of the rare earths have only one or two isotopes (Figure 1-3) whereas even numbered ones have many.

At the same time, nuclear stability rules also can explain why the abundances fall off with increasing Z. The anomalies normally arise in the process of nuclear synthesis, but calculations of astrophysicists, based on nuclear physics, permit explaining the excess of iron and the shortage of Li, Be, and B, not on their rates of formation but on their destruction by nuclear reactions after their formation.*

---

* For a discussion of the problems of nucleo-synthesis, see the excellent book by H. Reeves, *Evolution Stellaire et Nucleo Synthesis*, Gordon and Breach-Dunod, 1968.

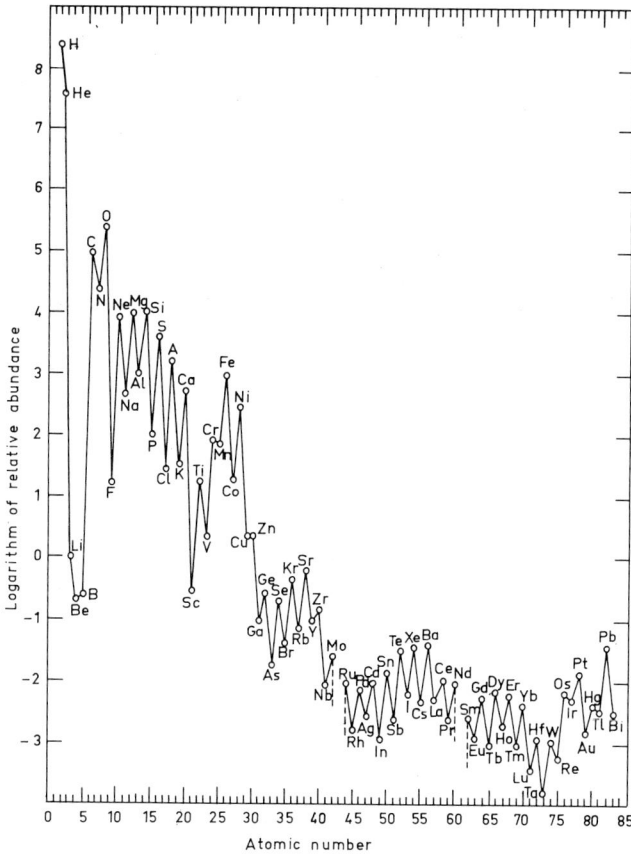

Fig. 1.2. Abundances of elements in the Universe as a function of atomic number (after Ahrens, 1965).

Returning to a point of view nearer to our objective for this book, we propose to compare the abundances found in the Sun with those found in meteorites of chondritic type (Figures 1-4 and 1-5). In this comparison, we exclude the gaseous elements, H, He, O, N, etc., which are much more abundant in the Sun for obvious reasons. An excellent correlation emerges between these two cosmic objects for all the elements except Ag, Li, and Hg.

The chondrites however appear to be cosmic objects that are relatively less fractionated chemically relative to original cosmic matter. Let us not forget that the Sun holds 99.8% of the mass of the solar system and that the meteorites probably came from the asteroid belt, situated between the orbits of Mars and Jupiter.

If one draws a similar abundance graph for the Earth's crust, its form is far less systematic, which discloses that the Earth's crust is a cosmic object that is highly differentiated from primeval matter. Making the same analysis on Moon rocks collected during the Apollo and the Luna missions, one must conclude that the lunar mares are more highly fractionated than the chondrites but much less so than the Earth's crust. To what can these differences be ascribed?

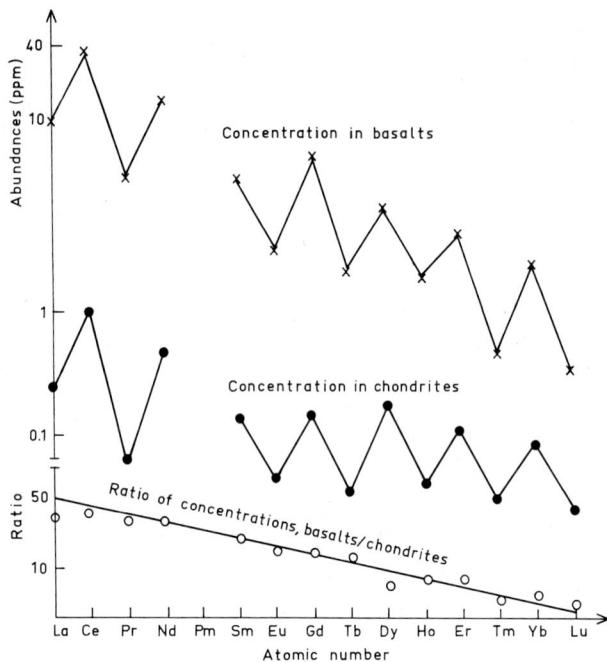

Fig. 1.3. Comparative abundances of lanthanides in basalts and in chondritic meteorites (after Schilling, Ph.D. Thesis, MIT, 1965).

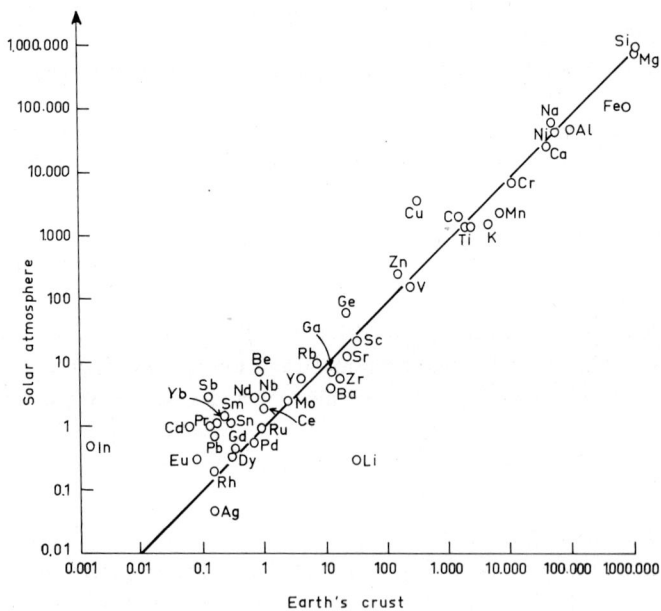

Fig. 1.4. Abundance of elements in chondritic meteorites and in the solar atmosphere (after Wood, 1968).

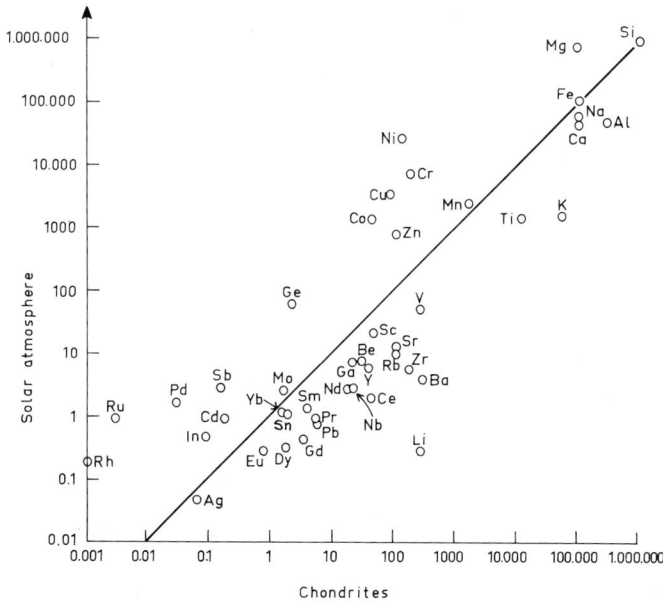

Fig. 1.5. Abundance of elements in the Earth's crust and in the solar atmosphere (after Wood, 1968).

A priori, two causes are conceivable: On the one hand, the chemical fractionation may be that prevailing at the time of formation of the planets (something that appears to have happened to the solar system approximately 4.6 billion years ago). On the other hand the local phenomena that occurred characteristically on each planet may ultimately have shaped each planet in its own pattern.

These ideas are well brought out by the following facts:

(1) The densities of the various planets, determined by astronomers, are not only not the same but vary widely (a fact normally attributed to varying fractions of Fe in the composition). Hence the total chemical composition of the planets is probably quite different.

(2) Astrophysical methods and more recent space flights disclose highly different surface conditions on the various planets. Thus the Moon has neither atmosphere nor water, Mars has an atmosphere and two polar caps composed of CO_2 and water, Venus has a dense atmosphere composed essentially of CO_2 at a very high temperature (450 °C), while the Earth has a gaseous atmosphere, flowing and circulating water, and two glacial polar caps.

The conditions for chemical fractionation are thus quite different on the various planets and so also should be the resulting observable surface features (rocks).

1.4. Chemical Compounds in the Solar System: Cosmological Minerals

Chemical elements combine to form compounds. What are the chief compounds of the solar system?

Leaving aside the inert rare gases, He, Ne, Ar, Kr, Xe, we find in succession

1.4.1. SIMPLE MOLECULES

These are H_2, H_2O, CO_2, N_2, O_2, CH_4, etc. They occur on the Earth; they appear on some of the planets where they constitute the atmosphere. Astronomers have detected some of them in distant stars, disclosing thereby a degree of uniformity in the structure of the Universe. In general, all are in the gaseous form, but they can be in solid or liquid form in special circumstances.

1.4.2. SIMPLE SOLIDS

These are in general the simple oxides: FeO, Fe_2O_3, TiO_2, or the sulfides: FeS_2, NiS, etc. They are found on Earth but also in meteorites and on the Moon.

1.4.3. SILICATES

These include all the minerals that arise from polymerization of the tertahedral SiO_4 radical, polymerizations that are electrostatically compensated by cations like Al, Ca, Mg, K, Na. The abundant silicates are limited in number and belong to the following families: quartz, feldspars, micas, amphiboles, pyroxenes, and olivines. They are found in terrestrial rocks, on the Moon, in meteorites, and they seem to make up the surface of all the nearer planets.

1.4.4. CARBON POLYMERS

These are the hydrocarbons and the components of living matter. They are found on the Earth, in some meteorites, and possibly on other planets. There are none on the Moon.

1.4.5. FE–NI ALLOYS

We shall reserve a special place for these composites for, while very rare on the surface of the Earth, they are an essential constituent of meteorites, and, recognizing their high mass density, they are assigned an important role in models of the internal composition of the planets. This also fits with the special place of these elements in the relative abundance curve of Figure 1-2.

The assortment of natural environments on the Earth and the planets is characterized by mixtures of these compounds. Such mixtures of solid compounds are called *rocks*. The variety of minerals is not infinite in number and only a few types are of importance in the Universe as a whole. Thus the surface of the planets is largely composed of silicate rocks. They may be surrounded by liquids or gases which flow over the rocks and which constitute the atmospheres (in a generalized sense).

The Fe–Ni alloys, important components of the Universe, appear to be confined to the cores of the planets.

1.5. Chemical Constitution of the Earth

The outer portion of the Earth accessible to geological examination consists of a series of successive 'envelopes' (Figure 1-6). The solid, topmost portion is called the lithosphere and it extends through the first approximately 100 km.

On this lithosphere rests the hydrosphere, comprising the totality of the water systems of the Earth: the seas, oceans, rivers, and glaciers. Not only does the hydrosphere cover the lithosphere over much of its area, in an obvious way, but it penetrates the surface to nearly 10 km.

At the surface of the lithosphere and hydrosphere exists the totality of life on Earth. The aggregate of living matter is called the biosphere.

Finally, everything is further surrounded by the gaseous envelope that is the atmosphere.

The composition and structure of these various envelopes are considered next.

1.5.1. THE ATMOSPHERE

From the surface of the Earth to a height of 60 km, the composition of the atmosphere is virtually uniform. It includes three major species, N_2, O_2, and Ar, and an array of minor ones: CO_2, Ne, He, CH_4, Kr, N_2O, H_2, O_3, Xe. At increasing distances from

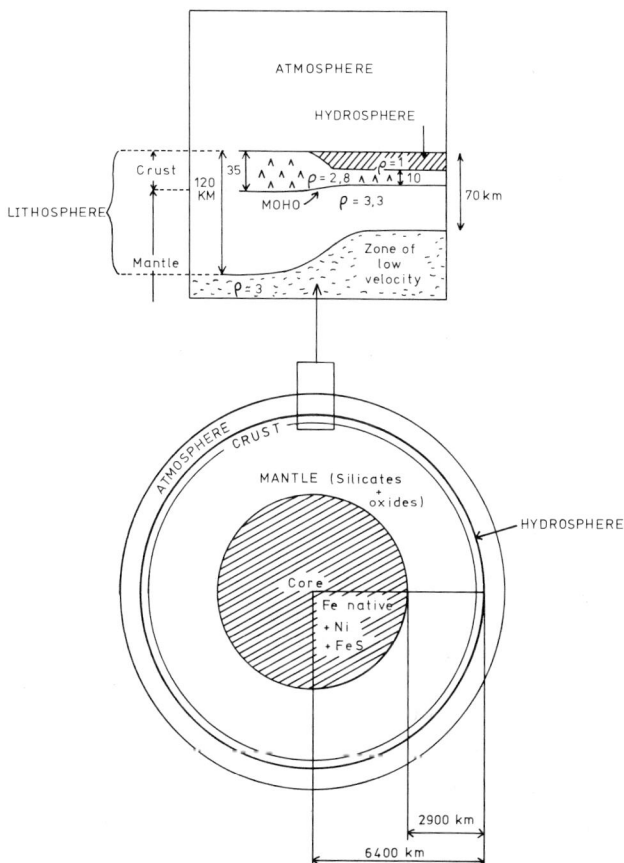

Fig. 1.6. Schematic diagram of the structure of the terrestrial globe.

the Earth, the atmosphere becomes attenuated, but even in the upper atmosphere, oxygen and nitrogen predominate.

Among the lesser constituents, two are highly important: ozone (O_3) which is particularly abundant in the stratosphere; it absorbs UV radiation. CO_2 is also an important gas particularly because of its role in the carbon cycle to which we shall return.

The total mass of the atmosphere is estimated to be 5.0×10^{21} g with minor variations according to various authors. This atmosphere is in motion (winds) and the motion is of major geochemical importance. Winds transport water vapor, clouds, and also fine solid particles. (Mid-ocean sediments, according to some authors, arise in part from this process.)

1.5.2. THE HYDROSPHERE

The oceans cover 3.6×10^8 km^2 of the Earth's surface, some 70% of the total. Assuming a mean depth of the oceans as 3800 m, this yields an ocean volume of 1.372×10^9 km^3. The total mass then becomes 1.413×10^{24} g according to Mason (1966). The balance of constituents in the hydrosphere is negligible compared with the oceans. The quantity of fresh water is estimated to be 0.5×10^{21} g and of ice, including the polar caps, is 22.8×10^{21} g. The mean composition of the hydrosphere is thus virtually that of the oceans.

The major ions dissolved in the ocean water are Cl^-, Na^+, $SO_4^=$, Mg^{++}. This composition will be examined in detail in Chapter 3. Like the atmosphere, the hydrosphere is affected by important currents and thus becomes a powerful agent for transport, whether of suspended or dissolved material.

1.5.3. THE BIOSPHERE

The significance of the biosphere in point of total weight is trifling compared with the other envelopes. Rankama and Sahama (1949) put the mass ratios of the hydrosphere, the atmosphere, and the biosphere at 70000:300:1. Qualitatively and dynamically, however, it is of enormous importance.

Riley has computed that the total production of organic carbon compounds on the Earth's surface is $1.46 \pm 0.87 \times 10^{11}$ t yr^{-1} of which $1.26 \pm 0.83 \times 10^{11}$ t are in the oceans and $2.0 \pm 0.5 \times 10^{10}$ t are on the continents.

If it is accepted that the biosphere has existed for 500 million years in a degree of development at least comparable with what we see today, the total mass that has existed in the biosphere exceeds the mass of the Earth!

The mean composition of the biosphere is hard to estimate as it is highly heterogeneous. Roughly, the biosphere is water (50% for the woods and 99% for the marine invertebrates) in which there are complex organic macromolecules. Five elements form the main mass of the biosphere, H, C, N, O, P, the others, while present, are quantitatively less important. The geochemical role of the biosphere will be found to be important in the balance of CO_2 and O_2 (photosynthesis, respiration), in the formation of skeletons and shells that ultimately provide important geological deposits (limestones, phosphates, etc.), and in the formation of organic sedimentary matter (coal, oil).

1.5.4. THE LITHOSPHERE

For many years, it was believed that the lithosphere was limited in depth, terminating at the seismic boundary called the Mohorovicic discontinuity (Moho). Recently, geophysicists specializing in plate tectonics have come to believe that the lithosphere extends much deeper. Acknowledging these divergent views, we shall speak initially of the crust as terminating at the Moho.

A fundamental problem then is the difference between the ocean floor and the continents. Under the ocean, the Moho lies at 15 km, under the continents at 35 km on the average. The transition from one value to the other occurs smoothly as has been shown in numerous studies carried out in Peru.

The ocean-floor crust consists of a thin skin of sediments covering a basaltic layer under which lies a layer whose nature is still poorly understood but is basic or strongly basic.

The continental crust is composed for the first 15 km of granites and gneiss (sedimentary rocks only constitute a small fraction). Deeper, there is a tendency to abandon the idea of a basaltic layer and instead to believe that the dominant composition must be a mixture of acidic and basic rocks* distributed in a heterogeneous way and corresponding to conditions of metamorphic facies involving high pressures and the absence of water. Such a combination is called a granulite assemblage.

Calculation of the mean composition of the crust began with Clarke and Washington (1924). On the basis of a series of laboratory analyses conducted by them, they published the first table of compositions. It showed among other things that 95% of crustal rocks are igneous or metamorphic. Poldervaert (1955) calculated mean crustal compositions taking into account the differences between the ocean floor and the continents and also data on the volume distribution in these domains. He found a composition fairly different from that of Clarke and Washington and probably more accurate. One can see from the table that Poldervaert's crust is more

	SiO_2	Al_2O_3	Fe_2O_3 + FeO	MgO	CaO	Na_2O	K_2O	TiO_2	P_2O_5
Clarke and Washington	59.12	15.82	6.99	3.30	3.07	2.05	3.93	0.79	0.22
Poldervaert	55.2	15.3	8.6	5.2	8.8	2.9	1.9	1.6	0.3

basic than that of Clarke and Washington. Eight elements (O, Si, Al, Fe, Ca, Na, K, and Mg) represent 99% by weight of the crust (the first three above account for 80%).

The amount of oxygen, already leading the elements in weight, leads even more

---

* The 'acidity' of a rock is tied to its silica content. Rocks called acid contain around 65%, those called basic contain about 50%.

markedly by volume, so that some writers have proposed naming the crust the 'oxysphere'.

Beside these major elements, there is a list of minor elements present in much lesser amounts, i.e., present in far less weight than the major elements. The mean values of their abundances are constantly being changed as analytical techniques advance (see the compilation in Table 1-I).

TABLE 1-I

Abundance of the elements in the Earth's crust (in g t^{-1})

1. H	1400	30. Zn	70	59. Pr	8.2
3. Li	20	31. Ga	15	60. Nd	28
4. Be	2.8	32. Ge	1.5	62. Sm	6.0
5. B	10	33. As	1.8	63. Eu	1.2
6. C	200	34. Se	0.05	64. Gd	5.4
7. N	20	35. Br	2.5	65. Tb	0.9
8. O	466000	37. Rb	90	66. Dy	3.0
9. F	625	38. Sr	375	67. Ho	1.2
11. Na	28300	39. Y	33	68. Er	2.8
12. Mg	20900	40. Zr	165	69. Tm	0.5
13. Al	81300	41. Nb	20	70. Yb	3.4
14. Si	277200	42. Mo	1.5	71. Lu	0.5
15. P	1050	44. Ru	0.01	72. Hf	3
16. S	260	45. Rh	0.005	73. Ta	2
17. Cl	130	46. Pd	0.01	74. W	1.5
19. K	25900	47. Ag	0.07	75. Re	0.001
20. Ca	36300	48. Cd	0.2	76. Os	0.005
21. Sc	22	49. In	0.1	77. Ir	0.001
22. Ti	4400	50. Sn	2	78. Pt	0.01
23. V	135	51. Sb	0.2	79. Au	0.004
24. Cr	100	52. Te	0.01	80. Hg	0.08
25. Mn	950	53. I	0.5	81. Tl	0.5
26. Fe	50000	55. Cs	3	82. Pb	13
27. Co	25	56. Ba	425	83. Bi	0.2
28. Ni	75	57. La	30	90. Th	7.2
29. Cu	55	58. Ce	60	92. U	1.8

Having reviewed the analyses of the various global envelopes, we are going to try to correlate the distribution of the elements with their electronic structure, that is, with their location in the periodic table of the elements.

1.6. Geochemical Classification (Goldschmidt, 1954)

Goldschmidt divided the elements into four main groups identified as siderophiles, chalcophiles, lithophiles, and atmophiles. This distinction is tied to the hypothesis of the geochemical subdivision of the planets into a dense core that is an alloy of Fe and Ni surrounded by an envelope of sulfides, then a layer of silicates, and finally the atmosphere.

In the core are concentrated certain elements of the Fe family and other transition elements whose abundance in higher layers is singularly weak (Figure 1-6). These elements form the family of siderophiles.

The family of chalcophiles includes on the one hand the elements whose properties border on those of sulfur: Se, As, Te,..., and on the other hand the elements that have a strong affinity for sulfur: Ag, Hg, Cu, Pb, Zn,....

The lithophiles include the elements that are concentrated in the Earth's crust. Among them are Si, Al, O, which form the foundation of the silicates and Na, Mg, Ca, Fe which enter into their structure.

Finally, the atmophiles include certain elements like H and N and the rare gases which 'are stored in the atmosphere'.

As can be found in Figure 1-7 this classification recurs in the periodic table of the

Legend:
Atmophile : N
Lithophile : Na
Chalcophile : Zn
Siderophile : Fe

H																	He
Li	Be											B	C	N	O	F	Ne
Na	Mg											Al	Si	P	S	Cl	A
K	Ca	Sc	Ti	V	Cr	Mn	Fe	Co	Ni	Cu	Zn	Ga	Ge	As	Se	Br	Kr
Rb	Sr	Y	Y	Nb	Mo		Ru	Rh	Pd	Ag	Cd	In	Sn	Sb	Te	I	Xe
Cs	Ba	La-Lu	Hf	Ta	W	Re	Os	Ir	Pt	Au	Hg	Tl	Pb	Bi			
			Th		U												

Fig. 1.7. Geochemical classification of the elements by Goldschmidt (1954).

elements. The lithophiles occupy the left and right sides of the table; the siderophiles and the chalcophiles are distributed among the central columns, and the atmophiles are on the extreme right.

More recently, various authors have proposed more complex classification schemes which however, preserve the main features of Goldschmidt's.

Finally it is necessary to bear in mind that certain elements are on the fringes of several groups. Thus Fe which is the basis of the siderophiles (sideros is the Greek word for iron) also plays an important role in the lithophiles.

1.7. The Earth as a Chemical Factory

All that we have just reviewed has been merely a description of the distribution of the elements in various portions of the Earth and has not been much concerned with why the distribution takes the form it does.

Now the Earth is a 'live' planet which changes continuously and which is the seat of great mechanisms that transport chemical elements and incorporate them into new and varied combinations (Figure 1-8).

The contact between the lithosphere and the hydrosphere is one of the most powerful sources of modifications of the Earth's surface. The water cycle (evaporation-precipitation-transport) literally washes down the continents. Water is a powerful corrosive agent; in flowing over the rocks, it charges itself with ions which it carries to the sea but in addition it breaks down the minerals leaving residual rocks that are named sand and soil according to their degree of development. Part of all this matter is mechanically transported by the water toward the ocean. In the ocean the detrital material settles to the bottom and some ionic reaction products precipitate.

Thus is produced the cycle of erosion and sedimentation which, if it were the only action, would transform the Earth into a smooth ball. All of the chemistry of the hydrosphere thus is dominated by its contact with the lithosphere, onto which in turn hangs the very existence of the biosphere.

In the lithosphere, two main types of phenomena originate. For one, the surface matter under certain conditions and at certain epochs undergo very considerable mineralogical transformations. These solid phase transformations occurred as a result of increased temperatures and pressures and are called metamorphic transformations. The net effects are essentially mineralogical and generally only in a relatively limited way chemical.

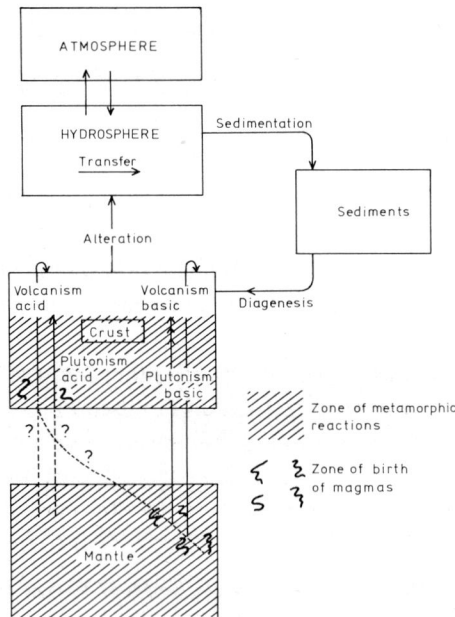

Fig. 1.8.   Diagram showing the dynamical relationship among the major chemical entities of the Earth.

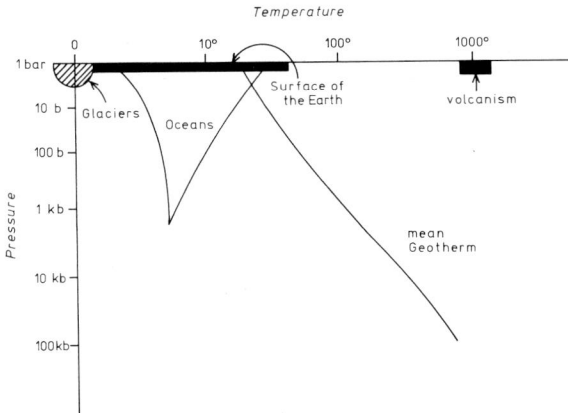

Fig. 1.9.   Pressure-temperature diagram showing the pertinent values for principal present day phenomena.

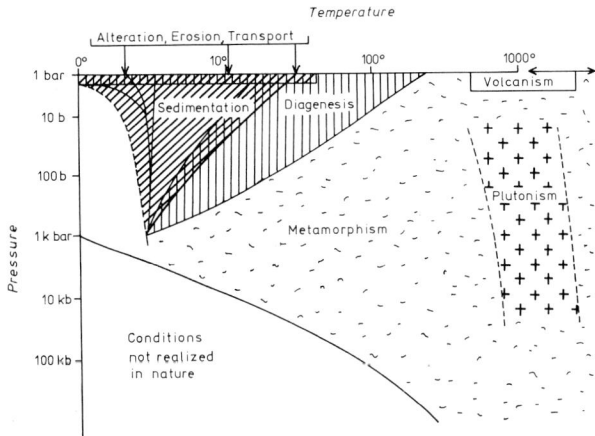

Fig. 1.10.   Pressure-temperature diagram showing the pertinent values for principal geochemical phenomena.

The other series of phenomena is much more important to the chemical fractionation of the globe. These are the magmatic phenomena. Pools of molten silicates formed at greater depths are squeezed out into higher levels of the crust where they freeze and crystallize. If they reach the surface, volcanism results; if confined to depths below the surface, the effect is called plutonism. How are magmas created and what is their composition? How did they evolve chemically? This is indeed a problem.

Geologists are beginning to realize that the chief characteristic of this chemical factory is that the various reactions taking place in it occur under widely varying conditions of temperature and pressure.

To fix our ideas since the beginning, we show in Figures 1-9 and 1-10 the various thermodynamic domains of the Earth and just where in the diagrams the great geochemical phenomena occurred.

One might propose to describe the geological cycles for each element in geo-chemical terms. This would lead to interesting generalizations but to relatively ele-mentary ones.

In this work, we shall attempt to disentangle certain main threads of chemical evolution. It is not our purpose to produce either a treatise or a catalog but to show by examples a line of reasoning.

References

Ahrens, T.: 1965, *Distribution of the Elements in Our Planet*, McGraw Hill, New York.
Goldschmidt, V. M.: 1954, *Geochemistry*, Oxford University Press, London.
Krauskopf, K.: 1967, *Introduction to Geochemistry*, McGraw-Hill, New York.
Mason, B. H.: 1966, *Principles of Geochemistry*, John Wiley and Sons, Inc., New York.
Miyake, Y.: 1955, *Elements of Geochemistry*, Maruzen Company, Tokyo.
Rankama, K. and Sahama, S.: 1949, *Geochemistry*, Univ. of Chicago Press, Chicago.
Shaw, D. M.: 1963, *Studies in Analytical Geochemistry*, Univ. of Toronto Press, Toronto.
Urey, H.: 1952, *The Planets, Their Origin and Development*, Yale Univ. Press, New Haven.
Wood, J.: 1968, *Meteorites and the Origin of the Solar System*, McGraw-Hill, New York.

CHAPTER 2

THE EQUILIBRIA OF PHASES IN THE LITHOSPHERE

The matter found on the Earth's surface can be divided into two categories: (1) that formed under the conditions of temperature and pressure prevailing at the surface, and (2) that formed at much higher pressure and temperature. Among the latter, two chief types may be distinguished:

The first is the magmatic rocks (also called igneous) that took shape from a liquid of melted silicates. The crystallization of these magmas could occur rapidly at the surface creating volcanic rocks or slowly at greater depths producing plutonic rocks.

The second is the metamorphic rocks that resulted from the recrystallization or the restructuring of rocks formed earlier by some other process not involving any melting. Thus the meta-igneous rocks are those formed from metamorphism of igneous rocks, the metasedimentary rocks are those metamorphosed from ancient sediments, and the polymetamorphic rocks are those that have undergone several successive metamorphisms.

The study of these two fundamental processes, the genesis of igneous rocks and the processes of metamorphism, has made considerable advances in recent years thanks to the thermodynamic approach based in particular on the idea of phase equilibrium. We are going to examine a few of these problems.

2.1. Metamorphic Equilibria

When a rock of given chemical composition is subjected to temperature and pressure different from those prevailing at its primary mineralogical assemblage, it is transformed into a new mineralogical assemblage. One says that it has undergone metamorphism.

Without offering the geological proofs of these processes, it may be said that modern research has demonstrated that one can identify most metamorphic rocks as assemblages of minerals that formed in the thermodynamic equilibrium conditions of P, T, and $P(H_2O)$ characteristic of metamorphism and remained unchanged since then.

Let us see several examples of these processes.

2.1.1. GENERAL CHARACTERISTICS OF METAMORPHIC TERRAINS

2.1.1.1. *The Paragenetic Association of Metamorphic Rocks*

The term paragenetic suite refers to an association of minerals that form in close contact in the same geological process so that each one affects the development of the others. In these terms, a metamorphic rock contains several paragenetic suites (Figure 2-1).

— the typical paragenetic association of metamorphism (one says typomorph) made up of minerals formed in the course of metamorphism;

    — the relic paragenetic association (or associations) formed before metamorphism

    — post-metamorphic paragenetic associations, formed since metamorphism.

Progress in the study of mineralogical structures by use of the polarizing microscope makes it possible today to identify objectively these three types of paragenetic suites in a metamorphic rock. We will only be concerned in what follows with the formation of the paragenetic typomorphs.

2.1.1.2. *Existence of Metamorphic Zones*

An inventory of these metamorphic paragenetic associations discloses that in any given type of rock, defined by its chemical composition, the number of paragenetic associations is relatively limited. In particular, the number is strongly limited for any given region. With this as a basis, one can mark off zones in which the paragenetic associations are constant (again, for a given chemical composition). These zones are separated by well-marked limits that coincide with the appearance or disappearance of one or several minerals.

It appears in addition that certain paragenetic suites associated with one chemical type of rock always occur together with certain other paragenetic suites associated with other chemical types of rocks. The inventory of this sort of coincidence of paragenetic suites demonstrates that the associations are not infinite in number and that one might catalog them in a relatively brief way. This gives what the petrographers call metamorphic facies. They are named on the basis of a given paragenetic associa-

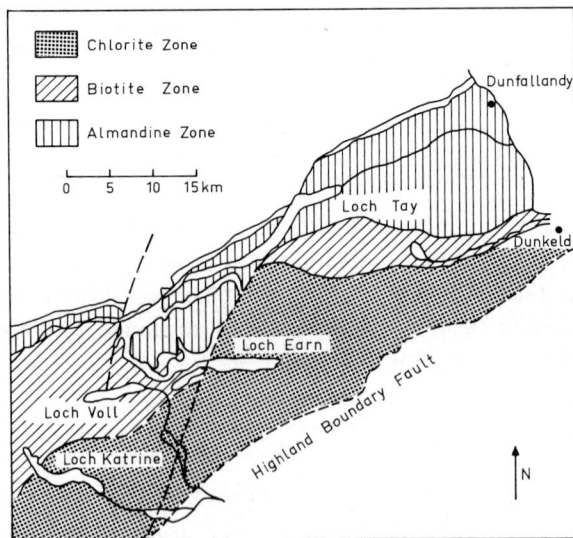

Fig. 2.1.   Zones of general metamorphism in the Dabradian of Scotland (after Tilley, 1962).

tion corresponding to a rock of given chemical composition. Thus to speak of the almandine-amphibolite facies does not signify that all paragenetic suites that are associated in that zone contain almandine but that the rocks having a 'basic' chemical composition contain almandine and amphibole.

In a given facies, there exists a one-to-one correspondence between chemical composition and paragenetic typomorph.

If one maps the metamorphic facies in a given area, one sees that their topologic relationship is well defined, that is to say, the facies may be recognized to follow one another in a certain order. The metamorphic facies thus look like assemblages of rocks having a special internal characteristic (one-to-one correspondence of paragenetic suite and chemical composition) and a relationship of order (Figure 2-2).

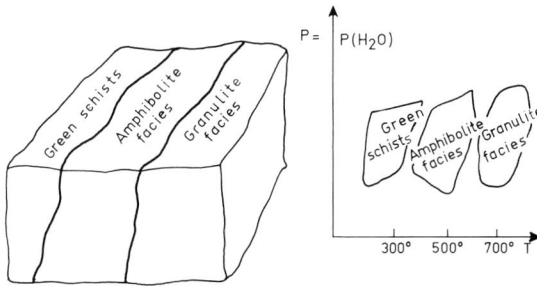

Fig. 2.2.   Relation between metamorphic facies and the P, T diagram.

2.1.2. THERMODYNAMIC INTERPRETATION OF METAMORPHIC FACIES

Metamorphic facies are created whenever nature subjects the rocks to conditions of temperature and pressure different from those that governed at their formation. Such conditions correspond to higher temperatures and pressures than those at the surface of the Earth. In the course of this phenomenon, supposed to be unique and called metamorphism, it is assumed that a new thermodynamic equilibrium is attained corresponding to the new thermodynamic conditions to which the rocks are subjected.

It is considered that the metamorphic paragenetic associations correspond to a new equilibrium state. The zones correspond to domains on the P, T, and $P(H_2O)$ diagram in which certain associations are stable. These zones are therefore two-dimensional, bivariant domains in the P, T plane. Their limits are the one-dimensional, univariant curves of the P, T plane.

Thus the relationship of order noted among the diverse metamorphic facies corresponds to the various stability domains of the paragenetic typomorphs on the P, T diagram (Figure 2-3).

As the equilibrium was supposed to become established, the one-to-one relationship

mineralogic composition $\rightleftharpoons$ chemical composition

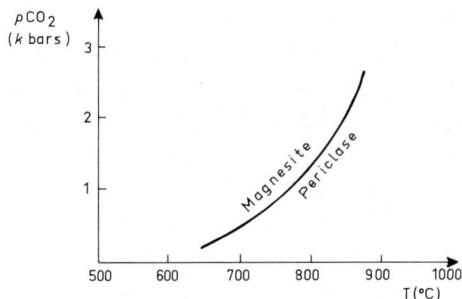

Fig. 2.3.   The polymorphic system Al_2SiO_5.

was preserved and the Gibbs phase rule $(V = C + 2 - \phi)^*$ was applicable; since $V \cong 2$, $C \cong 6$, ϕ must be ~ 6, which explains the small number of minerals that come into play.

The metamorphic terrains thus look like fossilized forms of the thermodynamic field that was prevailing at the time of metamorphism.

One of the goals of the study of metamorphic terrains is to learn the law of correspondence for each relationship:

chemical composition = paragenetic association

and the thermodynamic variables, pressure, temperature, pressure of H_2O, CO_2, etc.

2.1.3. Determination of metamorphic phase diagrams

The method used for this purpose is to make a laboratory study of phase diagrams controlled to correspond to the observations of the terrain. In practice, the stability of mineralogical assemblages and the corresponding equilibrium curves are studied in the course of experiments made under various conditions of temperature and pressure.

The simplest equilibrium curves theoretically are those that relate to polymorphic transitions. For instance, aluminum silicate has three polymorphs, kyanite, sillimanite, and andalusite. These three polymorphs have only one point in common on the P, T plane. In effect, Gibbs' rule, $V = C + 2 - \phi$, yields $V = 0$ since $C = 1$ and $\phi = 3$. In the same way, it can be shown that every mineral has two equilibrium domains (2 dimensional) which are separated from one another by a univariant curve.

The discovery of one of these polymorphs in a metamorphic equilibrium mixture permits at once establishing its approximate location on the P, T diagram if the experimental equilibrium curve is known.

Unfortunately, such polymorphic transitions are rare, and one is confronted with more complex systems in which several minerals occur.

An attempt is usually made to study monovariant equilibria in order to establish well defined limits. It is furthermore necessary that the reaction go sufficiently rapidly that true equilibrium is reached in the course of the experiment.

* V = variance; C = number of independent constituents; ϕ = number of phases.

We consider a classical example, the reaction

$$\underset{\text{Magnesite}}{Mg\,CO_3} \rightleftharpoons \underset{\text{Periclase}}{Mg\,O} + \underset{\text{gas}}{CO_2}$$

The Gibbs phase rule, $V = C + 2 - \phi$, becomes $2 + 2 - 3 = 1$. The equilibrium is thus univariant and is illustrated in Figure 2-4. The pressure of CO_2 is plotted against the temperature, and the univariance appears as a single curve along which magnesite and periclase coexist. Harker and Tuttle determined the equilibrium curve starting either from magnesite or from periclase in an atmosphere of CO_2.

By thus systematically studying univariant equilibria among minerals, it has been possible to establish a whole network of P, T curves showing metamorphic associations. We present by way of demonstration a P, T diagram due to Winkler (Figure 2-5). This network is undergoing continual improvement as new research develops and the example should only be considered as an illustration of one method.

It is necessary to make several comments on the uses of such mineralogical phase networks.

(a) In a given terrain, the existence of this or that mineral does not depend solely on the conditions of P and T that applied at the moment of the metamorphism but also on the chemical state of the rocks of the terrain. Thus one may find ideal conditions of P and T for the appearance of andalusite in a certain zone yet not find any of it because the Al content of the schists is inadequate.

(b) Experimental results as such are not the basic and fundamental constants of the processes. Beside the difficulty of measuring temperatures and pressures in high pressure experiments, one is never certain of having reached true metastable equilibrium in the laboratory. The times available for reaction are poor compared with those available to the minerals in nature. All this explains in part how the P, T diagrams happen to change with technical progress.

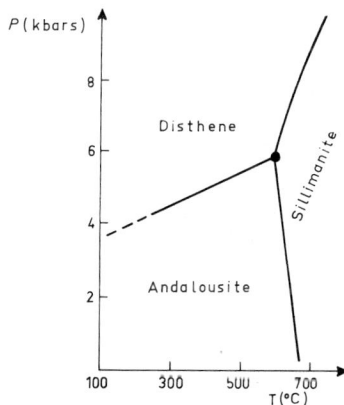

Fig. 2.4.   Graph showing the equilibrium of magnesite-periclase $+ CO_2$

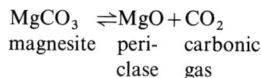

$$\underset{\text{magnesite}}{MgCO_3} \rightleftharpoons \underset{\substack{\text{peri-}\\\text{clase}}}{MgO} + \underset{\substack{\text{carbonic}\\\text{gas}}}{CO_2}$$

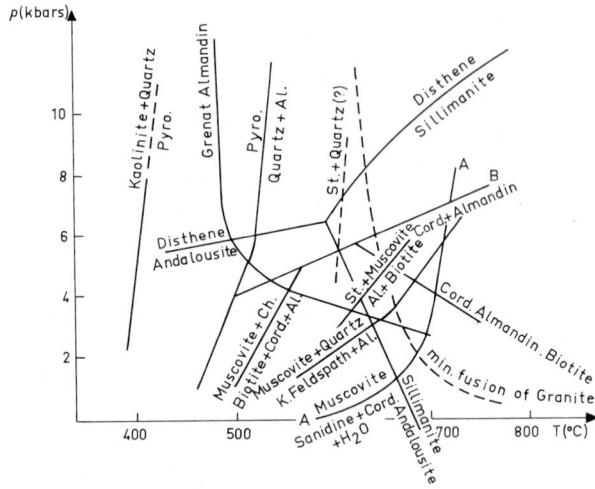

Fig. 2.5. Winkler's network (simplified). St., Staurolite; Ch., Chlorite; Al., Aluminum silicate; Cord., Cordierite; Pyro., Pyrophyllite.

(c) The fluids (CO_2, H_2O) that are present during metamorphism and which are found either in the minerals (biotite, amphibolite) or as fluid inclusions play a role that is not reflected in the P, T diagrams. In order to be strictly rigorous, it would be necessary to make a four-dimensional network showing simultaneously P, T, $P(H_2O)$, and $P(CO_2)$.

(d) In addition, real systems are chemically quite complex, and the mineralogical reactions that can occur interfere with one another. Certain reactions may follow one another sequentially, others may occur together and compete.

Nevertheless, the effort continues to develop a catalog of the various paragenetic suites of minerals obtained under various P, T conditions and with various chemical compositions.

2.1.4. CONCLUSIONS. GENERAL CHARACTERISTICS OF METAMORPHISM

Metamorphic reactions are chemical reactions that occur among solid minerals in the presence of interstitial fluids (H_2O, CO_2) that doubtless serve to transport ions in the reactions. These reactions are caused by the progressive changes in temperature and pressure. As these external conditions change, the system responds with mineralogical reactions that restore the system to thermodynamic equilibrium.

Thermodynamics has thus enabled us to explain, starting from the mineral combinations we find on the surface of the globe, how we might reconstruct the conditions that existed during metamorphic periods very long ago.

In all the foregoing material, we have assumed that metamorphism consisted of a series of reactions occurring with only small disturbances from equilibrium. This model of the equilibrium among minerals, which has to its credit having stimulated most of the decisive advances of these last 20 yr in the study of metamorphic phenomena does not get by without posing several problems of principle.

In the foregoing, we have assumed that the history of the rock could be divided into three periods: the premetamorphic period, the metamorphic period, and the postmetamorphic period. The conditions of metamorphism are assumed to be such that (1) the mineralogic reactions that occur in it virtually totally blot out the pre-existing paragenetic associations; (2) they create new paragenetic associations that reflect the equilibrium conditions of the metamorphism; (3) finally that these paragenetic associations are preserved from destruction during the postmetamorphic episodes.

There is here a phenomenon of recording that is extremely selective in the time and the intensity of the phenomena that it records. It can be understood by reference to kinetic phenomena and barrier potentials, recognizing that metamorphic reactions begin at $300\,°C$ for at this temperature the thermal energy $(\frac{3}{2}kT)$ first exceeds the energy of activation for the reaction of minerals like muscovite or pyrophylite.

Conversely, it must be recognized that the temperature of metamorphism drops abruptly after its climax, permitting the protection of the assemblages of typomorphs of the metamorphism by a tempering phenomenon. This double limit permits (on the whole) a recording of events of higher temperature at a given site.

As a matter of fact, without reopening the main threads of this reasoning, the modern studies are going to improve the scheme considerably by showing:

(1) that it is necessary to speak of numerous metamorphisms for a given region and not of just one; polymetamorphism; and

(2) that in numerous cases the reactions proceed according to non-equilibrium laws.

The first point emerges from the body of this work but we shall only attack the second point when taking up irreversible processes.

2.2. Phase Equilibria in the Silicate Magmas

Without dwelling for the moment on how magma is created, let us picture the chemical behavior of a magma of molten silicate that on freezing in the crust undergoes fractional crystallization. To fix our ideas, especially since we shall be dealing with the vast majority of terrestrial magmas, we shall examine essentially those magmas that lead to the granites and those that lead to the basalts.

2.2.1. BASALTS AND GRANITES

If one makes a statistical analysis of the silica (SiO_2) content of igneous rocks outcropping at the Earth's surface, the resulting histogram displays two maxima, at 65% and at 53%. Petrographic examinations correlated with this histogram show that two well known types of rocks are associated with the two maxima: granites and basalts. One model of the Earth's continental crust, for many years the classical model, proposed that under a layer of 15 km of granite was a deeper layer of basalt that in turn was separated from the mantle by the seismic discontinuity called the Mohorovicic.

Considering surface rocks only, granites (and gneisses) are the most abundant rocks at continents' surface and conversely basalt is the major component of the ocean floor. The observations just cited reveal the importance of the basalts and the granites in the overall composition of the crust and what geochemical interests hinge on their mode of creation.

A granite is composed of quartz, feldspars, and often of ferromagnesian minerals (biotite, hornblende, hypersthene) or white mica (muscovite). To these major minerals accessory minerals are added, such as zircon, apatite, and more rarely sphene (titanite), etc. Although variable, the structure is usually granular, that is, the major minerals are present in crystals of macroscopic size. Generally, the crystals are not preferentially oriented and yield a sort of fluid-like structure. However, when tectonic phenomena have influenced a granitic mass after its deposition, the crystals show preferred orientations correlated with the constraining fields of force imposed on the rock.

Basalts by contrast are volcanic rocks composed of a paste of microscopic crystals and of phenocrysts that 'swim' in this paste. Actual mineralogical analysis is difficult and is replaced by virtual mineralogical analysis based on chemical analysis. Such an analysis is called normative. Petrographers have shown that in the case of basalts the norm and the mode (actual analysis) differ only slightly. The mineralogical composition of basalts is a mixture of pyroxenes and plagioclases to which may occasionally be added either nepheline, olivine, or by contrast a little tridymite (in the normative form, generally). These rocks all arise from the crystallization of a pool of molten silicates called a magma. Geochemical study of the origin of these rocks is going to provide us the occasion for showing the manner of discussing magmatic phenomena while simultaneously trying to resolve some important geochemical problems.

2.2.2. THE GENESIS OF GRANITE. EXPERIMENTAL RESULTS. STUDY OF PHASE DIAGRAMS

We begin with a rapid scan of the systems albite-anorthite, albite-orthoclase, quartz-orthoclase, and quartz-orthoclase-albite.

2.2.2.1. *The Albite-Anorthite System*

The plagioclases form a continuous solid solution extending from pure albite, Si_3AlO_8 Na at one extreme to pure anorthite, $Si_2Al_2O_8$ Ca at the other. The thermodynamic diagram of this system includes three domains separated by two boundaries (Figure 2-6a):

(1) a single-phase domain of magma toward high temperatures.

(2) a single-phase domain of solid plagioclases toward low temperatures.

(3) Between these two, a two-phase domain comprising both magma and plagioclase at the same time.

Domain (3) is bounded by two curves that come together at a point on the pure-albite axis at one end and at a point on the pure-anorthite axis at the other end, the ordinates (T) of the end points being the melting temperatures of the pure substances respectively. The two curves are known as the solidus and the liquidus.

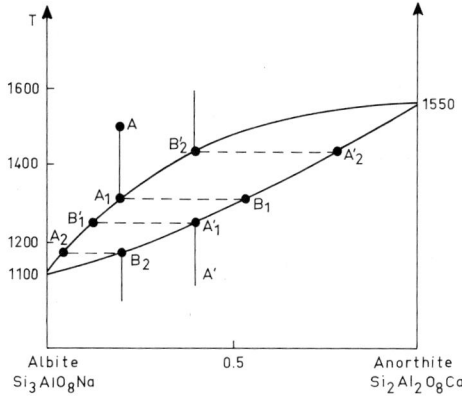

Fig. 2.6a.   Phase diagram for the system albite-anorthite.

Let us see how to use such a diagram. Suppose we have a magma represented by point A in Figure 2-6a (the composition is fixed by the abscissa of point A) and chill the magma until it is frozen. Between A and A_1 nothing special happens. At A_1, a plagioclase starts to crystallize, and its composition is the abscissa value for point B_1. The remaining magma grows poorer in anorthite, its representation migrating along the liquidus to point A_2, and during the cooling crystallizing out more plagioclase whose composition is migrating toward the abscissa value of B_2. The process continues in this way until the magma is completely gone. At each instant, the composition of the phases is controlled by the instantaneous value of the temperature and the application of this procedure.

Following complete crystallization, the system continues to cool without further change of composition.

On the whole, the law of crystallization is recorded in the structure of the plagioclase which begins with a core rich in anorthite and builds outward with a series of coronas progressively richer in albite.

This mechanism is at the root of the well known observations of zoning of plagioclases (in detail, the exact phenomenon of zoning is more complicated because of the occurrence of diffusion).

If the law of variation of temperature with time is investigated, it has the appearance shown in Figure 2-6b. It can be seen that the step on the curve is not perfectly level. This is due to the existence of latent heat of crystallization that slows the cooling process.

Next let us consider the inverse process of partial fusion or melting. We start with

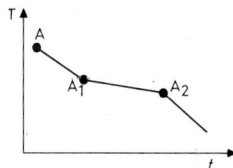

Fig. 2.6b.   Evolution of temperature during the course of time in the albite-anorthite system.

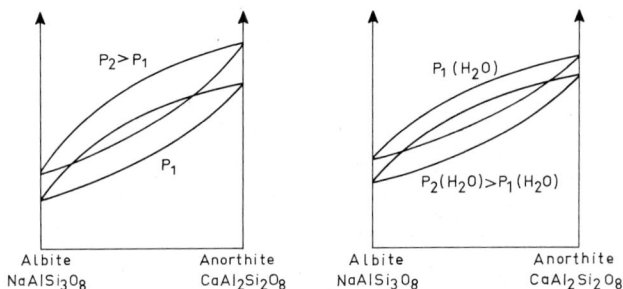

Fig. 2.7.   Influence of pressure and of water pressure on the albite-anorthite diagram.

a solid of composition represented by A′ in Figure 2-6a and heat it. When point A′$_1$ is reached, a melt of composition B′$_1$ begins to appear. The farther the melting progresses, the nearer the composition of the melt approaches A′ and the richer the remaining solid grows in anorthite. The influence of different pressures on such a graph is shown in Figure 2-7. It is clear that this influence is different depending on the presence or absence of water.

2.2.2.2. *Quartz-Potash Feldspar System*

This phase diagram (Figure 2-8a) is very different from the preceding one because the two components are immiscible and do not form a solid solution even slightly. The solidus is horizontal and the liquidus is broken into two clearly distinct parts. Let us see how such a diagram 'works'.

Suppose that a magma represented initially by point A is progressively chilled. When it reaches the liquidus at point B, it begins to deposit pure potash feldspar and progresses along the liquidus toward C. When C is reached (the eutectic point) a mixture of crystals of quartz and feldspar precipitates until the melt is used up.

If we had started from a magma with the composition (abscissa value) of point A′, upon cooling to the liquidus it would have begun to precipitate silica at once, then after reaching point C, a mixture of silica and feldspar. The mixture precipitating at C has definite proportions of silica and feldspar called the eutectic mixture.

In the present case, the cooling curve differs from the previous one (Figure 2-8b). After a rate of cooling of the melt dictated by the conductivity of the magmatic mass, from A to B, the cooling becomes slower from B to C because of the latent heat of crystallization of the feldspar. Upon reaching C, the temperature becomes fixed and remains so until all the liquid has vanished. At the eutectic temperature there is thus a level step in the cooling curve. Thereafter, the solid continues to cool normally.

Now let us consider the inverse phenomenon of partial fusion. We begin by warming a solid with composition indicated by point A$_1$ in Figure 2-8a. When the temperature reaches the eutectic temperature *T*, liquid starts to appear. This liquid has the eutectic composition (point C). On further heating, the liquid follows the liquidus curve until it reaches point A′$_1$.

It is clear that regardless of the initial composition of the solid, the first liquid to appear always has the same composition, that of the eutectic mixture.

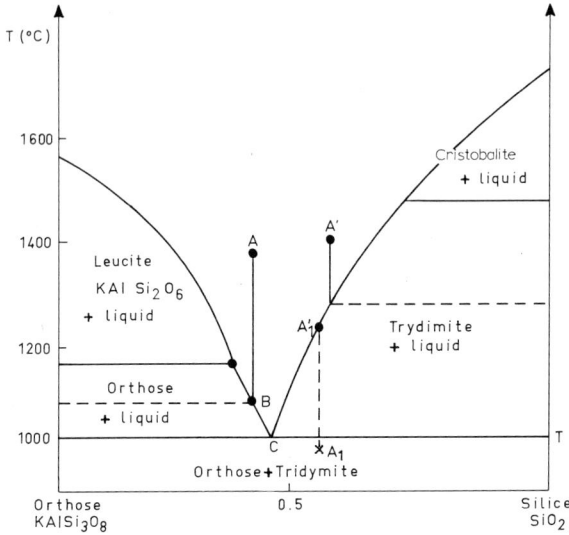

Fig. 2.8a.   Phase diagram for the system orthoclase-silica.

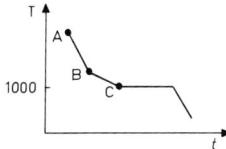

Fig. 2.8b.   Evolution of temperature with time in the orthoclase-silica system.

2.2.2.3. The Albite-Potash Feldspar System

In this case, we will speak only of the liquid-solid equilibrium mixture, omitting the intermixing of the solids after crystallization. Figure 2-9 is the associated phase diagram.

A magma is again represented by point A and is cooled progressively. Upon cooling to point B, a solid crystallizes out of the magma with the composition of point C, that is to say, a solid solution of albite and potash feldspar. The magma, on further cooling, then progresses to B', the solid to C'. When the magma reaches D, it crystallizes to completion to a solid solution with composition D. (The term azeotrope is applied to a mixture that does not change its composition on changing phase.)

Again let us examine the inverse procedure and heat a solid initially at point A' on the diagram of Figure 2-9. On reaching B'₁, the solid starts to melt, the molten material having the composition of C'₁ that is ordinarily not the composition D of the azeotrope but is narrowly determined by A'. On further heating, the magma develops along the liquidus moving away from the azeotrope.

It thus becomes apparent that there is a marked distinction between a eutectic and an azeotrope. For one, the initial composition of the melt is fixed, regardless of the composition of the solid; for the other, the composition is variable.

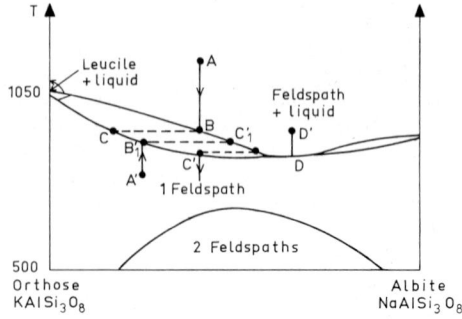

Fig. 2.9.   Phase diagram for the alkaline feldspars in the presence of water $p(H_2O) = 2000$ bar).

2.2.2.4. *Complex Two-Component Systems*

(a) Thermal barriers. The diagram (Figure 2-10) may be thought of as the combination of two simple eutectics. The two vertical lines corresponding to the E_1 and E_2 eutectic compositions form a barrier in fractional crystallization phenomena. The same is true for the vertical line defined by the pole XX'. Fractional crystallization will not allow liquid composition to migrate across this 'barrier'.

(b) Peritectic relations. In a two component system the proportion of phases in equilibrium is given by the lever rule. For example in the binary eutectic (Figure 2-10) the relations $m_{C_I} \cdot \overline{BA} = m_{liq} \cdot \overline{CA}$ holds, where m_{C_I} is the amount of component I, and m_{liq} is the amount of liquid with composition C.

It is possible that eutectic temperature variation with pressure causes an eutectic to overpass a temperature maximum (thermal barrier). Such is the case for the system leucite – silica (Figure 2-11). At high pressures the diagram is similar to (Figure 2-10). At low pressures the virtual leucite-K-feldspar eutectic lies on the same side of the thermal barrier as the quartz-K-feldspar eutectic. For a liquid composition intermediate between leucite and K-feldspar (point A) we may write:

$$m_{leucite} \cdot \overline{AG} = m_{liquid} \cdot \overline{Az},$$

z is the point on the liquidus corresponding to a given temperature.

The relation holds down to a temperature corresponding to D. At E we may write:

$$m_{leucite} \cdot \overline{AG} = m_{K\text{-}feldspar} \cdot \overline{AF}.$$

However, since $\overline{AF} < \overline{AD}$ we have that $m_{K\text{-}feldspar} > m_{liquid}$. For this to be true we must admit that a reaction

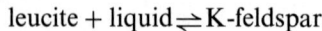

$$\text{leucite} + \text{liquid} \rightleftharpoons \text{K-feldspar}$$

takes place.

This reaction principle was shown by Bowen to be of great importance in the genesis of magmatic rocks.

2.2.2.5. *Total Pressure and Fluid* (H_2O) *Pressure*

Figures 2-7, 2-10, and 2-11 show how pressure can affect the geometric relations in

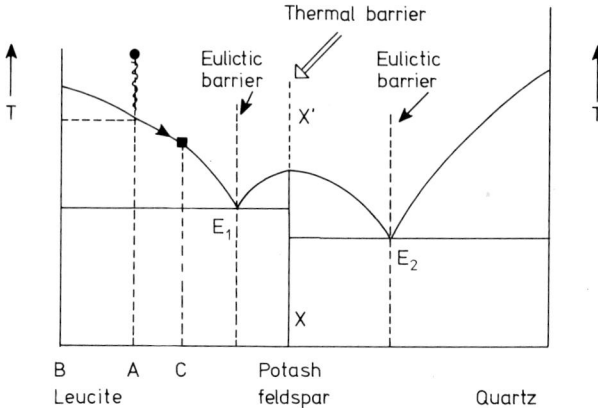

Fig. 2.10.   Phase diagram for leucite-potash feldspar-quartz at high pressure.

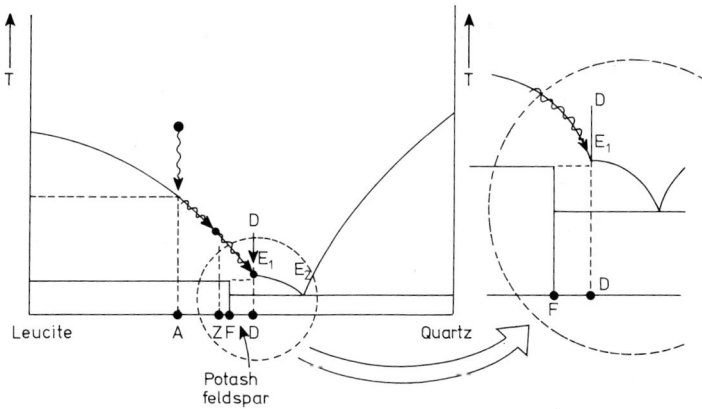

Fig. 2.11.   Phase diagram for leucite-potash feldspar-quartz at low pressure.

phase diagrams. In general an increase in 'dry' total pressure enlarges the area where
solids are stabile, whereas an increase in fluid (H_2O) pressure enlarges the area of the
liquid. Eutectic compositions may vary greatly with total and/or fluid pressure often
in non simple fashion.

Binary systems show many geometrical relations that can also vary with pressure.
Certainly natural rock systems cannot be reduced to binary and are complex multi
component systems. It is, however, sometimes possible to limit the discussion of rock
genesis to relatively simple ternary systems. Our purpose here is to illustrate the
rationale of granitic and basaltic rock origin in the light of three component systems
with variable total and fluid pressure.

2.2.2.6. *The Ternary System Silica-Albite-Orthoclase*

Tuttle and Bowen have determined the phase diagram (Q-Ab-Orth) in the presence
of an *excess of water* (Figure 2-12). Their results show that no ternary eutectic point

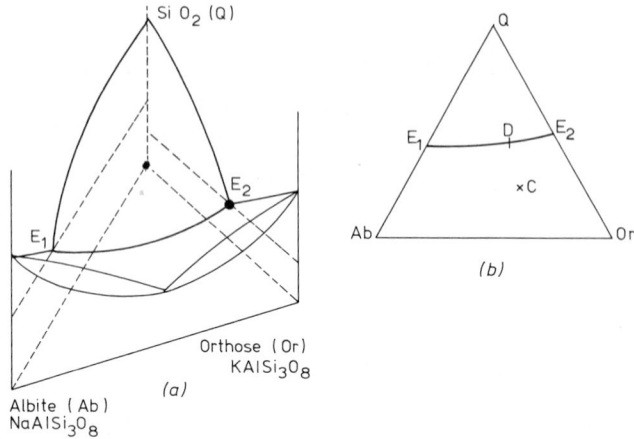

Fig. 2.12a–b.   Silica-orthose-albite system.

exists but that there is a cotectic line joining the two binary eutectic points. Thanks to this diagram, the microscopic textures observed in granite can be understood.

When a magma has a composition that falls into the region QE_1E_2 of Figure 2-12a the first mineral to crystallize out is the quartz, then, when the melt reaches the cotectic line, it crystallizes into a mixture of silica and alkaline feldspars (which are themselves a solid solution that separates under certain conditions).

If the initial magma has composition C, the alkaline feldspar crystallizes out first; then after the melt reaches 'point' D, a mixture of quartz and alkaline feldspar crystallizes.

It should be noted that the composition of the quartz-feldspar mixture depends on the initial composition of the magma (as well as the corresponding composition of alkaline feldspar).

This model of crystallization is well verified by the observations of petrographers which show that when the composition of a granite lies in the zone E_1E_2 Ab Or of Figure 2-13 the texture shows large crystals of automorphic feldspar surrounded by a mixture of small crystals of quartz and feldspar. When the composition corresponds to the region QE_1E_2 in the figure, it is the quartz that is automorphic surrounded by a mixture of quartz and feldspar.

Since the observed facts are compatible with the experimental assumptions, we can conclude with Tuttle and Bowen that the model for formation of granite by crystallization from a magma is coherent.

On the other hand, if we envisage the phenomenon of partial fusion of granite, we conclude because of the absence of ternary eutectic that the first formed melts will have highly variable compositions tied to the composition of the initial material. Considering the extreme uniformity of the composition of granites, Tuttle and Bowen inferred that there is no magmatic recycling of granites. The granites, they say, must be derived from basalts by fractional crystallization.

Winkler (1965) and his associates reperformed the experiments of Tuttle and Bowen and brought out a fundamental fact. Despite the small trace of calcic plagioclase, this substance has a decisive influence on the diagram. Its presence even in trace amounts causes a ternary eutectic minimum to appear in the preceding diagram. The minimum temperature depends on the content of anorthite. This author concluded that a rock having a granitic composition (granite, gneiss, or sedimentary rock of greywacke type) undergoing limited partial fusion leads to a granitic magmatic melt (anatexis).

The experiments of both research groups were done with excess water present. The temperature of fusion and crystallization is therefore in the range of 650 to 750 °C; *if the water is absent fusion does not set in below the* 900 *to* 1000 °C *temperature range.*

2.2.2.7. *Comparison of Experiments with Field Observations*

Granites are found at the Earth's surface principally in two forms:
– A massive form with sharp contacts against the surrounding sedimentary material (type 1);
– A complex form with diffuse boundaries changing gradually into high grade metamorphic assemblages (type 2).

Bowen's hypothesis can explain granite of type 1 but does rather badly for type 2. In addition, even for the granites of type 1, one rarely sees the great masses of basaltic rock that ought to be associated.

Winkler's hypothesis explains granites of type 2 very well for one can see in the field the gradual transition starting from non-molten metamorphic rocks, then partially molten rocks (migatites), and finally the totally molten rocks without directional fabric (granites). For granites of type 1, Winkler calls upon the concept of thermal imbalance. If the temperature rises in a heterogeneous metamorphic complex, certain parts melt first under 'eutectic conditions', others remain unmolten. If the temperature continues to rise, the partially molten portions that hold at the fixed eutectic temperature rapidly find themselves in density unbalance with their surroundings and tend to float toward the crustal surface and to 'inject' themselves into the upper layers giving birth to granites of type 1.

The explanation of granites of type 2 in this manner fails to explain an extremely general field relationship, the presence of basic rocks (gabbros and diorites) that are associated with these massifs. Bowen's theory naturally handles these facts much better.

In conclusion then, the magmatic origin of granites is beyond doubt, anatexis seems equally well established, but the origin of granites of type 2 remains debatable.

This discussion goes far beyond the limits of a simple difference of thermodynamic interpretations and brings out, in fact, an important problem of the chemistry of our planet.

Granite exists only on the continents of the Earth, true granite being absent from the moon and meteorites. It seems to be a characteristic of the Earth.*

* Recently, the studies from Venera 8 by gamma spectrometry of U, Th, and K have suggested that granites can also be present on Venus.

If the interpretation by Winkler is correct, it is possible that granites are formed from initial sediments which were themselves formed from the cycle of erosion and sedimentation. This last process, seemingly one unique to the Earth, gives a clue to why granite is limited to the Earth. If Bowen's views are right, it is difficult to understand why granites are not found in other parts of the universe. (Perhaps even this argument will fall with further space exploration.)

Still, if Bowen is right, one must admit that the continents formed at the expense· of the mantle by magmatic differentiation. On the other hand, if the Winkler theory applies, a more complex process involving sedimentation must be conceived. Here then one must wait patiently for further scientific progress and avoid dogmatic excesses.

2.2.3. THE GENESIS OF BASALTS

Contrary to the situation for granites, the magmatic origin of basalts does not pose a problem for anyone since one can see basalt crystallizing on the surface of the earth from melts (volcanic lava flows). The problem that does arise is rather that of the origin of the magma, of the nature of its precursor, and of the mechanisms that explain the assorted chemistry observed. It may be said that no complete theory of the origin and evolution of basalts enjoys universal acceptance. We must content ourselves with unfolding the directions taken by some of the studies.

2.2.3.1. *The Evolution of Basalts and Fractional Crystallization*

Let us consider first the phase diagram for the system forsterite-quartz-anorthite (Figure 2-14) and discuss with its aid the evolution of a basaltic magma. As to composition, the various basaltic rocks have roughly the composition of the cross-hatched portion of Figure 2-13. Suppose for example that a magma has the composition of point A. At first, a little olivine (labeled forsterite in Figure 2-13) will crystallize out, then rejoining the equilibrium line between pyroxene and olivine, pyroxene begins to crystallize with the olivine (point B). The process continues to point C where one begins to see concurrent crystallization of olivine, pyroxene, and plagioclase (anorthite in Figure 2-13). After reaching point C, the evolution exhibits a special phenomenon. The olivine reacts with the melt to give a pyroxene and the proportion of that mineral grows. At D where the quartz begins to precipitate, the olivine has vanished completely. Because of the requirement of mass balance, the amount of quartz is small by comparison with the initial amount of magma.

If the initial magma had the composition shown by point X, the first mineral to crystallize would be pyroxene, followed by plagioclase (anorthite). If Y were the starting point, anorthite would crystallize first, olivine (forsterite) second, then pyroxene, etc.

Despite its simplicity the diagram allows exposition of some extremely general observations:

(1) It is quite easy to explain the relation that exists in nature between the order of crystallization of minerals, basalts, and their chemical compositions.

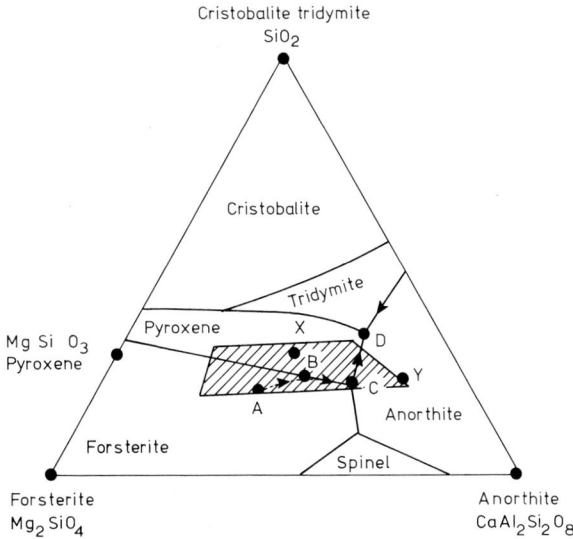

Fig. 2.13.   Silica-Forsterite-anorthite system.

(2) The diagram exposes a certain very important process that occurs between the melt and the crystals that have already precipitated, whereby these crystals redissolve and allow precipitation of other minerals. Bowen has shown that in addition to the relation

$$\text{olivine} + \text{magma} \rightarrow \text{pyroxene}$$

we can also get

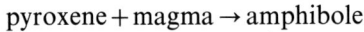

$$\text{pyroxene} + \text{magma} \rightarrow \text{amphibole}$$

and

$$\text{amphibole} + \text{magma} \rightarrow \text{biotite}.$$

(3) If in the course of crystallization one raises by any means whatever (in nature, this could be a volcanic eruption) a portion of the magma that has not yet crystallized, one gets a magma progressively more and more acid ending ultimately with a highly siliceous one. A simple picture thus emerges of what may happen in a magmatic reservoir where, by fractional crystallization, the magma may grow more and more acid, i.e., highly siliceous.

(4) If a solid consisting essentially of olivine and pyroxene (peridotite) is warmed, the first liquid to appear will have a composition of pyroxene and plagioclase, hence essentially a basaltic one. It is seen there the principle of the genesis of a basalt from an ultrabasic starting point.

In fact, to study real cases, the diagram has to be made more complicated:

(a) The olivine in general is not forsterite but a solid solution fayalite-forsterite (Fe_2SiO_4-Mg_2SiO_4).

(b) The plagioclase is generally a solid solution anorthite-albite.

(c) The pyroxenes are numerous and complex and cannot be reduced to enstatite.

(d) For phenomena occurring at depths (crystallization in a magmatic chamber or partial fusion of deep-lying portions of the lithosphere) the pressure can seriously modify the diagram.

Having hastily shown the principle of these studies of differentiation of basaltic magmas, we now turn to examine the origin of the magmas.

2.2.3.2. *The Origin of Basaltic Magma*

Very careful studies of American volcanoes in Hawaii and Soviet ones on Kamchatka have shown that basaltic magmas originate at a depth of 60 to 90 km. This zone is definitely beneath the Moho discontinuity of seismologists and therefore lies in the upper mantle.

Information concerning the composition of the upper mantle derives from four types of investigations:

(a) Seismology discloses that the speed of P waves below the Moho discontinuity is 8.2 km s^{-1} and as a consequence that the density of the material at this level is in the neighborhood of 3.2 to 3.3 g cm^{-3}. This value already limits the number of minerals which are candidates for forming the upper mantle.

(b) The material contained in basalts and satisfying the preceding conditions is of two types: peridotites, with or without garnet (olivine plus a little pyroxene) and eclogites (sodic pyroxene plus garnet (pyrope)). Both of these combinations have densities around 3.3 g cm^{-3}.

(c) The complete or partial melting of the material of the upper mantle must give rise to basalt. This requirement places strict limits on the chemical composition of the mantle which must necessarily be basic or ultrabasic.

(d) The mineralogical assemblages satisfying the preceding conditions must be stable under the conditions of temperature and pressure insofar as they may be estimated from the extrapolation of the known heat flow at the surface.

Practically, the assemblages, i.e., the rocks that satisfy these conditions are essentially peridotites with or without garnet, eclogites, and a rock conceived by Ringwood, pyrolite (composed of pyroxene and olivine) or else pyrolite with garnets.

The hypothesis of eclogite has against it the following fact: The composition of eclogite is identical with that of basalts, however, if the mantle is eclogitic, it follows that volcanism must be a process of total fusion, and this is hard to accept for various thermodynamic and mechanical reasons (Ringwood and Green, 1966). By means of chemical analysis and by the use of calculus of probabilities, it is however possible to know the theoretical mineralogical composition of a basalt.

This last varies considerably. To represent the various varieties of basalts, Yoder and Tilley (1962) proposed a tetrahedron (Figure 2-14) at the apex of which are located the principal minerals (normatives) of basalts. The tetrahedron is divided into three domains representing the alkaline basalts or basanite, whose virtual composition may be written (Neph + Cpx + Plag + Ol), basalts with olivine (Oli + Opx + Cpx + Plag), and quartz tholeites (Quartz + Plag + Opx + Cpx).

This tetrahedron once more uniquely used here for the nomenclature was studied

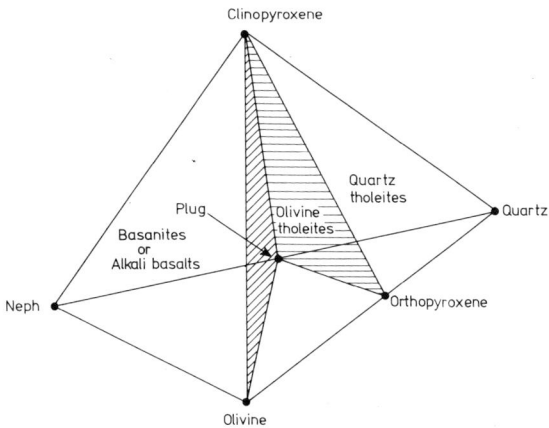

Fig. 2.14.   Yoder and Tilley Tetrahedron simplified.

by Yoder and Tilley (1962) and others from the point of view of phases at low and high stresses. These authors have shown that the planes of separation of Cp Plag Ol and Cp Plag Opx were thermal barriers. If one recollects the example of Figure 2-10 one can see that it is impossible to 'cross' the thermal barrier by fractional crystallization and to pass from one side of the barrier to the other. It will be the same even if the system is infinitely more complex. It thus follows that the single fractional crystallization of such a magma cannot lead to formation at one and the same time of quartz tholeites, olivine basalts, and alkaline basalts.

Submarine exploration coupled with more and more systematic study of recent volcanism on the globe shows that the rocks with tholeitic affinity on the one hand and with alkaline affinity on the other constitute the two most abundant classes of basalts. From this arises the problem of how to conceive their relationship. Do they arise from two different types of mantle? But if so, how should the coexistence of the two types on a worldwide scale be explained? Do they arise from the same source, but if so, by what mechanism?

These are the questions of a great debate in the course of which the most redoubtable polemic talents of the petrologists of the world have clashed – without reaching agreement.

Some basalts erupt in the oceans on the midoceanic ridges. As a consequence, thanks to the phenomenon of sea floor spreading, they form the upper layer of the oceanic crust. It is primarily tholeitic basalts that emerge on the ridges. Alkaline basalts are found in the oceans only on sea-mounts and in volcanic islands (Hawaii, the Canary Islands, Cape Verde Islands, etc.).

On the continents, eruptions of basalts occur in the vicinity of great fissures. These are chiefly tholeitic basalts (western United States, Deccan). Basaltic volcanoes also appear in the centers of continents, along great faults (Central Massif (France), African rift). These are alkaline basalts. Finally, the active belts of the Pacific have numerous basaltic volcanoes despite the fact that the dominating volcanism of these regions is andesitic.   All this is of course much oversimplified.

One of the important problems of contemporary petrology is to understand why in some cases tholeitic basalts are formed and in others alkaline basalts.

2.2.3.3. *Hawaii and the Model of G. A. MacDonald*

In the Hawaiian Islands tholeitic and alkaline basalts are observed at the same volcano.

Through a fairly systematic study, it appears that the tholeitic basalts with olivine were the earlier ones emitted and that the alkaline basalts came later.

G. A. MacDonald admits that the tholeitic basalts with olivine would be the first magma to issue directly from the mantle. Then, after pausing in a magmatic reservoir, this basalt would become differentiated, precipitating olivine and pyroxenes. These dense minerals would migrate toward the bottom of the reservoir while inversely, the alkalines, especially sodium, would migrate toward the top. This last mechanism of magmatic fluids can be seen in the lava lakes of Hawaii.

These two mechanisms would thus give birth to the tholeitic and the alkaline basalts (Table 2-I).

TABLE 2-I

Analysis of an alkaline basalt and of a tholeite
(Engel *et al.*, 1965; Cann, 1971)

	Basalt	Tholeite
SiO_2	48.16	49.61
TiO_2	2.91	1.43
Al_2O_3	18.31	16.01
Fe_2O_3	4.24	11.49
FeO	5.89	
MnO	0.16	0.18
MgO	4.87	1.84
CaO	8.79	11.32
Na_2O	4.05	2.76
K_2O	1.69	0.22
P_2O_5	0.93	0.14

2.2.3.4. *The Rittman-Kuno Relationship*

In the volcanic arcs of Indonesia and Japan, Rittman and later Kuno disclosed an important relationship. On the oceanic side, the basalts are tholeitic, and the farther one advances toward the continent, the more the alkaline character is accented (Figure 2-15a, b).

Within the framework of the theory of sea floor spreading, island arcs are interpreted as arising from the subduction of lithosphere plates created at the ridge. It is proposed that partial melting of the plate occurs along the length of the subduction zone (Benioff zone). The chemical differences observed at the surface are due only to the progressive influence of a growing pressure on the composition of the magmatic melts.

Ringwood and his colleagues have supported this hypothesis with partial fusion

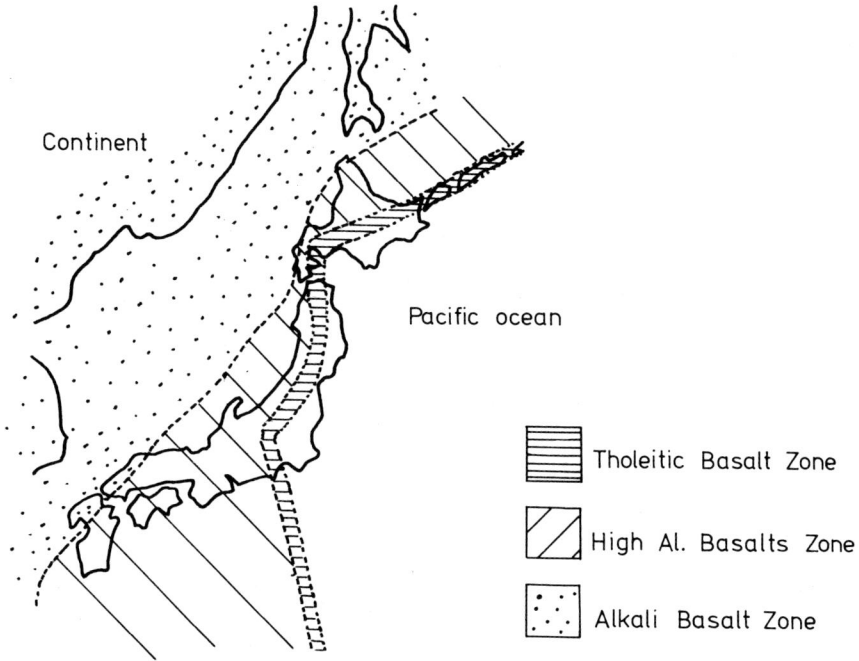

Fig. 2.15a.   Position of tholeites and alkaline basalts in a geodynamic island arc system. (Japan arc system).

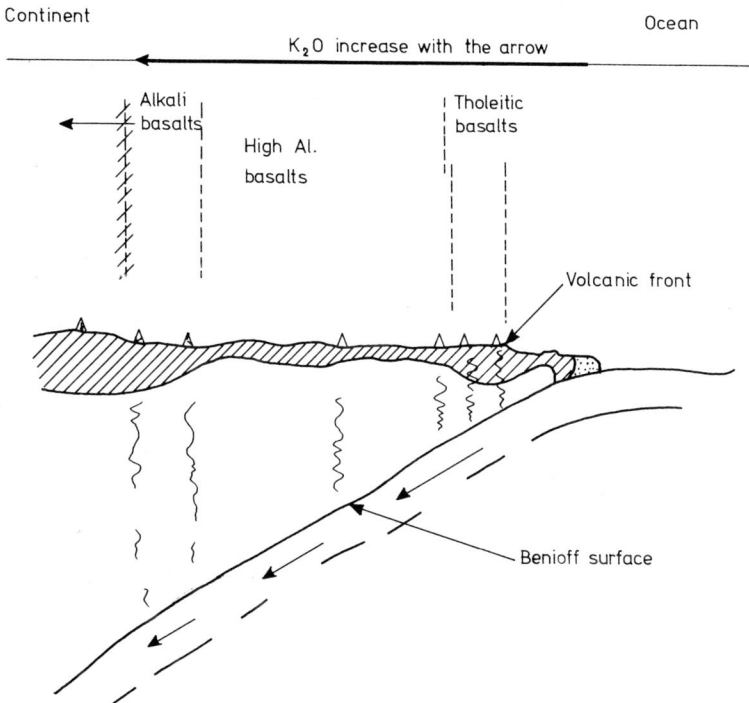

Fig. 2.15b.   Cross section of Japan Island showing the zoning in Basalt Type distribution.

experiments conducted at high pressure. They showed that starting from a solid of pyrolitic type, basaltic melts could be produced by partial fusion. The higher the pressure the less the degree of partial fusion and the richer the melt becomes in alkalines. The lower the pressure the greater the melting and the more important the content of SiO_2.

Tholeites would thus form from an advanced fusion of the mantle at low pressure while alkaline basalts would form at greater depth from a more partial fusion.

In these two examples which lead to two radically opposing theories, it is important to distinguish the facts underlying the interpretations.

The relation between the alkaline character of basalts and the distance from the trench of an island arc appears to be general and is to be included under the heading of facts (however such a relation may be used, for example, in paleographic reconstitutions.).

The interpretation by contrast is far from being unique, and it is objective to say that today no theory enjoys universal acceptance even though the scientific community leans with the prevailing fashion alternatively toward one or the other.

One can imagine that the choice in reality is also naive and that one needs to take account of numerous phenomena that are not considered at all in these explanations, such as the role of gases, the migrations of volatile matter, the reactions with the reservoir, etc.... This is a chapter that will doubtless see profound evolution in a future approach.

In summary basalts result from partial fusion of the upper mantle, of ultrabasic or basic composition. After the formation of the magma, this material can stop during its rise toward the surface and in the resulting magmatic chamber, a differentiation can take place.

Today we tend to explain all these processes with the aid of a phase diagram of the minerals, but it is probable that magmatic gases play an important part.

2.3. Magmatism or Metamorphism. The Conditions for Their Appearance

When a rock at the surface sinks, or more generally, when the values of P and T of its environment increase, what happens then?

Suppose the temperature rises, while the pressure rises concurrently. At first, in the range from 250 to 700 °C, a series of metamorphic transitions in the solid state will occur (to be sure, most such effects depend on water as the carrying agent that causes the small displacements of matter necessary for these transformations).

Near to 700° various possibilities arise:

(a) For rocks with basic or ultrabasic composition, new metamorphic transitions are produced which cause the appearance of minerals like garnet-pyrope, sodic pyroxene (omphazite), etc.

(b) For rocks of acid composition in the presence of water, melting begins giving rise to a granite melt (anatexis).

(c) For rocks of acid composition but without water, metamorphism continues

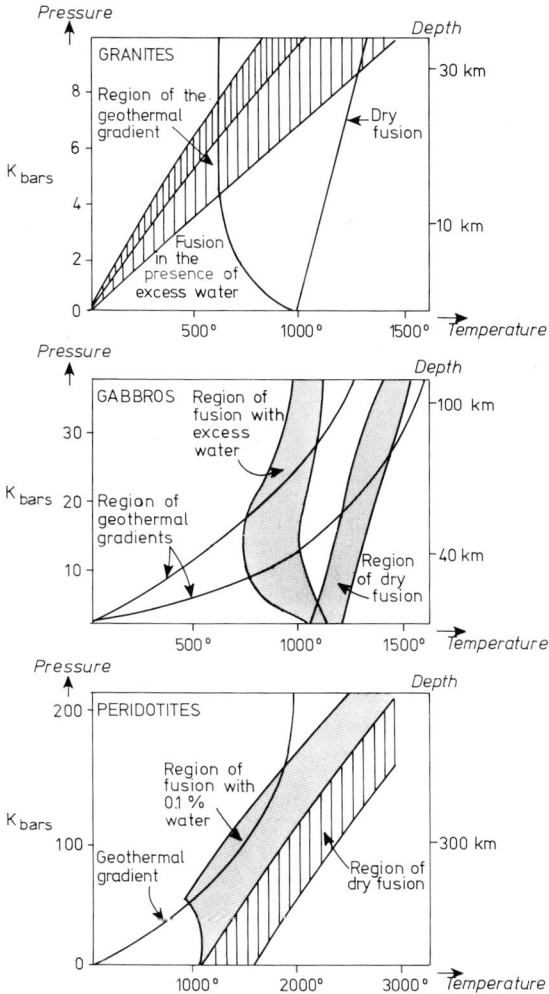

Fig. 2.16.   Pressure-temperature diagrams showing fusion curves for granites, gabbros, and peridotites with or without water (after Wylie, 1968).

leading to associations like

> potash feldspars – plagioclase
> pyroxene – garnet, etc.?

These associations are known by the name of granulite facies.

Suppose the temperature rises to 1200°.

Ultrabasic rock, if it is there in the presence of water, melts. By contrast, in the absence of water, it must be heated to 1350 or 1400° in order to melt, something that doesn't occur in 'normal' conditions in the Earth. (All the reasoning is based on assuming that the variation of temperature as a function of pressure is normal.)

By contrast again, as soon as one passes through the zone where the temperature is 1600°, that is, through a depth of 350 km, the effect of pressure is such that only a considerable increase in temperature (and one with little likelihood) can bring about fusion of the silicates. All reactions from this depth onward are solid state reactions. This description is shown graphically in Figures 2-16.

The schematic reasoning has as its goal to show that under the admissible conditions of the Earth:

(a) The zones where the magmas are created are localized at certain domains of depth.

(b) The fusion depths are a function of the chemical composition and in particular of the water content. This last declines statistically from the surface downward but this decline has a 'geography' that has a considerable influence on the mechanism of what occurs.

References

Bowen, N. L.: 1928, *The Evolution of Igneous Rocks*, Princeton Univ. Press, Princeton, N.J.

Cann, J. R.: 1971, *Phil. Trans. Roy. Soc. London* **A268**, 495.

Engel, A. E. J., Engel, C. J., and Havens, R. C.: 1965, *Bull. G.S.A.* **16**, 719.

Gorshof, G. S.: 1970, *Volcanism and the Upper Mantle* (translated from Russian), Plenum Press.

Kuno, H.: 1959, *Bull. Volcanologique* **20**, 37.

Poldervaert, A. and Hess, H. H.: 1967, *Basalts*, Vols. 1 and 2, Wiley-Interscience, New York.

Ringwood, A. E.: 1962, *J. Geophys. Res.* **67**, 4005.

Ringwood, A. E. and Green, D. H.: 1966, *Tetonophysics* **3**, 383.

Turner, F. J.: 1968, *Metamorphic Petrology*, McGraw-Hill Publ. Co., New York.

Turner, F. J. and Verhoogen, J.: 1960, *Igneous and Metamorphic Petrology*, McGraw-Hill Publ. Co., New York.

Winkler, H. G. F.: 1965, *Petrogenesis of Metamorphic Rocks*, Springer-Verlag, Berlin.

Wyllie, P. J.: 1968, *Ultramafic and Related Rocks*, Wiley-Interscience, New York.

Yoder, H. S. and Tilley, C. E.: 1962, *J. Petrology* **3**, 342.

CHAPTER 3

CHEMICAL EQUILIBRIA IN THE HYDROSPHERE

At the Earth's surface, at temperatures below $50\,°C$, reactions among solids proceed at extremely slow rates. Only reactions in solutions can go at a reasonable speed and reach equilibrium.

The entire surface of the globe, previously designated the hydrosphere, is dominated by chemical reactions in aqueous solutions. In this chapter, an overall view will first be given of reactions in the hydrosphere, then an example of broad general interest – the saturation of the oceans with $CaCO_3$.

3.1. Chemical Reactions in the Hydrosphere

3.1.1. CLASSIFICATION OF REACTIONS

Natural water carries in solution essentially inorganic species that are more or less dissociated into ions. The reactions among all the dissolved substances are very numerous. Accordingly, it has been attempted to classify them with the object of restricting the mass of information needed to take adequate cognizance of all of them.

The logical expositions of this classification are numerous and are designed to include extension to non-aqueous solvents (Charlot, Trémillon). We shall content ourselves here with recalling the chief results that are of most particular interest to geochemistry.

Accordingly, we shall start with homogeneous reactions in solutions and those reactions that cause interaction of the solution with some other phase, solid or gaseous. For the geochemist, these last are the interactions of the hydrosphere with the lithosphere or atmosphere. While this chapter is limited to these processes, they are in fact the only ones that apply to the global cycle of the elements and thus they are of prime interest to us. Refreshing our recollection, we note that if ions of type A and type B react to form the insoluble precipitate AB, the law of mass action may be written

$$AB \underset{\downarrow}{\rightleftharpoons} A + B$$

and the solubility product is*

$$(A)\,(B) = K_s.$$

This relationship is true only if the solution is in contact with the precipitated solid AB. If it should not be, the equation changes to the inequality

$$(A)\,(B) < K_s.$$

* Parentheses denote activities; brackets denote concentrations.

As to single phase (homogeneous) reactions, they may be regarded as an exchange of particles (p) between two pairs called donors and acceptors (A_1, D_1 and A_2, D_2):

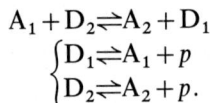

$$A_1 + D_2 \rightleftharpoons A_2 + D_1$$
$$\begin{cases} D_1 \rightleftharpoons A_1 + p \\ D_2 \rightleftharpoons A_2 + p. \end{cases}$$

If the exchanged particle is a proton H^+ the reaction is called an acid-base reaction. The donor is called acid, the acceptor, base.

If the exchanged particle is an electron e^-, we are dealing with an oxidation-reduction reaction. The donor is the reducing agent; the acceptor is the oxidant or oxidizing agent.

In the more general case, it is customary to speak respectively of the reaction forming a complex, of the complex, and of the complex ion.

Every donor-acceptor pair is characterized by an equilibrium constant:

$$\frac{(A_1)\,(p)}{(D_1)} = K_1.$$

Water intervenes in acid-base reactions as it is at one and the same time both acid and base. Thus as an acid it gives an H^+

$$H_2O \rightleftharpoons OH^- + H^+$$

and as a base it accepts an H^+

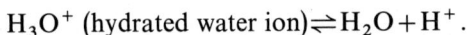

$$H_3O^+ \text{ (hydrated water ion)} \rightleftharpoons H_2O + H^+.$$

If acids, symbolized by AH, become completely dissociated in water,

$$AH + H_2O \rightleftharpoons A^- + H_3O^+,$$

they are called strong acids. The corresponding negative ion A^- loses its basic character.

By comparison, bases stronger than OH^- cannot exist in water. Their corresponding acids no longer possess their acid-like properties and are called inactive.

In an analogous way, water is an oxidant;

$$H_2O + e^- \rightleftharpoons \tfrac{1}{2}H_2 + OH^-$$

and a reducing agent,

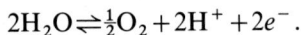

$$2H_2O \rightleftharpoons \tfrac{1}{2}O_2 + 2H^+ + 2e^-.$$

Reducing agents stronger than hydrogen cannot exist in water (example, Na), the corresponding oxidized form being inactive in water (example, Na^+). Neither do oxidants stronger than oxygen (F_2) exist in aqueous solution, and the corresponding reduced forms (F^-) are inactive in water.

In a solution where acid-base reactions occur, or where complex reactions occur, the activities of the hydrated H^+ ions and the complexing ions lead to a characterization of the solution as to acidity or degree of complexing. In particular, knowledge of the state of the solution gives the information needed for knowing the degree of dissociation of all the corresponding acceptor-donor systems.

For reasons of numerical convenience, a logarithmic scale has been adopted whereby one writes:

$$pH = -\log(H^+)$$
$$pe = -\log(e^-)$$
$$pL = -\log(L),$$

where L denotes the complexing particle.

3.1.2. ACID-BASE REACTIONS IN NATURAL WATER

3.1.2.1. *Principal Factors*

Among the ions present in significant quantities in natural water (see Table 3-I), the

TABLE 3-I

Concentrations (molalities) of the elements in natural waters

Element	Probable form in sea water	Sea water	River water (mean value)
Cl	Cl^-	0.5459	0.19×10^{-3}
Na	Na^+	0.4680	0.26×10^{-3}
Mg	Mg^{2+}	0.546×10^{-1}	0.16×10^{-3}
S	$SO_4^=, MgSO_4^\circ, CaSO_4^\circ$	0.282×10^{-1}	0.12×10^{-3}
Ca	Ca^{2+}	0.103×10^{-1}	0.44×10^{-3}
K	K^+	0.99×10^{-2}	0.56×10^{-4}
C	HCO_3^-	0.23×10^{-2}	0.77×10^{-3}
Br	Br^-	0.84×10^{-3}	0.17×10^{-6}
B	H_3BO_3	0.44×10^{-3}	0.18×10^{-5}
Si	SiO_2	0.11×10^{-3}	0.60×10^{-3}
Sr	Sr^{2+}	0.91×10^{-4}	0.65×10^{-6}
F	F^-	0.68×10^{-4}	0.10×10^{-4}
Li	Li^+	0.29×10^{-4}	0.23×10^{-6}
P	$HPO_4^=, H_2PO_4^-$	0.22×10^{-5}	0.63×10^{-6}
Rb	Rb^+	0.14×10^{-5}	0.16×10^{-7}
I	I^-	0.39×10^{-6}	0.16×10^{-7}
Zn	Zn^{2+}	0.15×10^{-6}	0.15×10^{-6}
Mo	MoO_4^{2-}	0.13×10^{-6}	0.52×10^{-8}
Fe	$Fe(OH)_2^+, Fe(OH)_4^- (?)$	0.12×10^{-6}	0.41×10^{-5}
Ba	Ba^{2+}	0.95×10^{-7}	0.29×10^{-6}
Ni	Ni^{2+}	0.85×10^{-7}	0.68×10^{-7}
Se	$SeO_4^=$	0.51×10^{-7}	?
As	$HAsO_4^=, H_2AsO_4^-$	0.40×10^{-7}	0.27×10^{-7}
V	$VO_2(OH)_3^=$	0.39×10^{-7}	0.20×10^{-7}
Mn	Mn^{2+}	0.37×10^{-7}	0.16×10^{-6}
Al	?	0.35×10^{-7}	0.11×10^{-4}
Cu	Cu^{2+}	0.31×10^{-7}	0.11×10^{-6}
Ti	?	0.21×10^{-7}	?
Co	Co^{2+}	0.20×10^{-7}	?
U	$UO_2(CO_3)_3^{4-}$	0.13×10^{-7}	?

only ones that have acid-base properties are the 'carbonates', the borates, and silica. The large values of pH of the boric and silicic acids and their relativity weak concentrations force them into a relatively secondary role. In particular circumstances, other systems may play a primordial role, these are $Fe^{3+}/Fe(OH)^{2+}/Fe(OH)_3$ in the wash waters of sulfurous minerals, or H_2S/HS^- and NH_4^+/NH_3 in reducing environments. But, the principal acid-base system of natural water is the system $H_2CO_3/HCO_3^-/CO_3^{2-}$.

The evaluation of pH thus falls back on a calculation 'of pH of a weakly acid solution' partially neutralized. Its degree of dissociation must be compatible with electrical neutrality. The equation of neutrality brings into play all of the ions present in the solution:

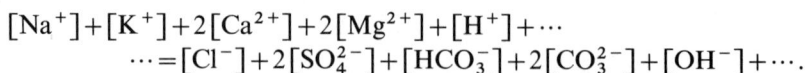

$$[Na^+]+[K^+]+2[Ca^{2+}]+2[Mg^{2+}]+[H^+]+\cdots$$
$$\cdots=[Cl^-]+2[SO_4^{2-}]+[HCO_3^-]+2[CO_3^{2-}]+[OH^-]+\cdots.$$

We can collect the terms corresponding to ions without acid-base characteristics:

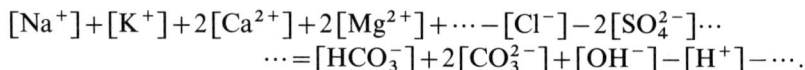

$$[Na^+]+[K^+]+2[Ca^{2+}]+2[Mg^{2+}]+\cdots-[Cl^-]-2[SO_4^{2-}]\cdots$$
$$\cdots=[HCO_3^-]+2[CO_3^{2-}]+[OH^-]-[H^+]-\cdots.$$

The ions on the left side only enter into their global charge contribution and they are assigned the name alkalinity, R_B.

An initial study of acid-base reactions in natural waters can be reduced to two factors, the carbonate system and the alkalinity. Let us then examine the limits of variation allowable to these factors and to what constraints they may be subject.

3.1.2.2. *The Carbonate System*

The various forms in solution, CO_2, H_2CO_3, HCO_3^-, and CO_3^{2-} are all correlated with the quantity of CO_2 in the gaseous phase. Two fundamental cases may be distinguished according to whether the natural water is or is not in contact with a gaseous phase.

(1) For surface waters, the atmosphere serves as an 'infinite' reservoir at constant pressure (p of $CO_2 = 3 \times 10^{-4}$ atm). Although the exchange of CO_2 between the atmosphere and the hydrosphere is not rapid, displacements from equilibrium of more than 10% do not last for a long time, and since the pressure of CO_2 enters only in logarithmic form, such a deviation is of little importance in the long run.

Water in the soil is in contact with an atmosphere greatly enriched in CO_2 by the phenomena of respiration of plants and animals. The pressure of CO_2 varies relatively little around the greatly elevated mean value of the order of 0.1 atm.

At constant CO_2 pressure, the curve of R_B vs pH takes the form shown in Figure 3-1. It should be noted in the same figure that the quantity $\partial R_B/\partial pH$, representing the buffering power (T) of the water, takes on significant values only at extremes of pH.

(2) In ground waters, and in deep, poorly agitated basins of waters, there is no exchange of CO_2 with the atmosphere, and the ultimate variations in the state of

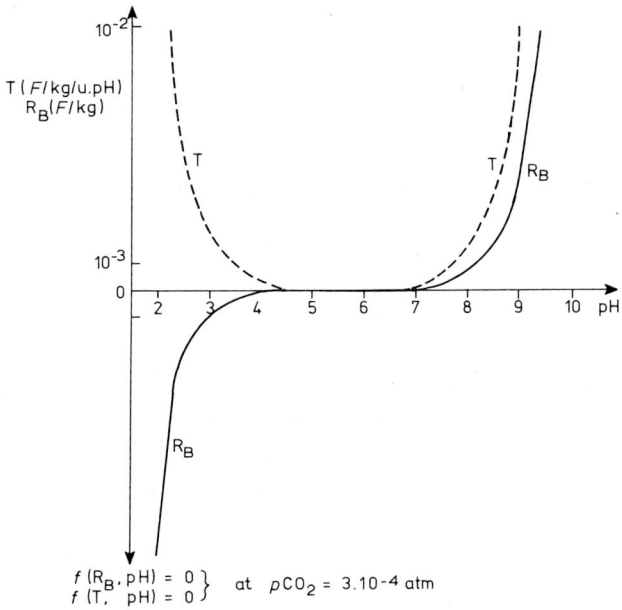

Fig. 3.1. Relation between pH and alkalinity at constant pCO_2.

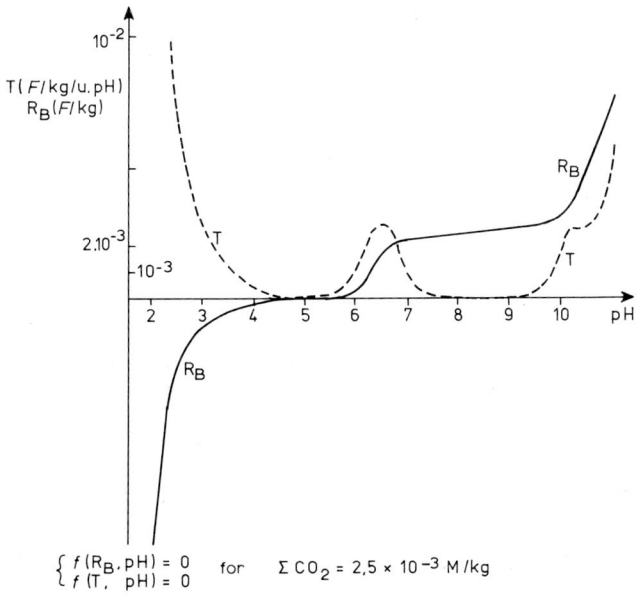

Fig. 3.2. Relation between pH and alkalinity at constant ΣCO_2.

dissociation of the various ionized species are limited by the total constant quantity of dissolved carbonates. ($\Sigma \, CO_2 = [H_2CO_3] + [HCO_3^-] + [CO_3^{2-}] = $ constant.)

Under these conditions, the relationship between alkalinity and pH as well as that between buffering power and pH (Figure 3-2) has a very different appearance. The frequency of occurrence of very slight acidity (pH between 6 and 7) may be increased there.

It should be noted finally that the foregoing relationship may be slightly modified quantitatively by variations in temperature or in ionic strength in the surrounding medium that may in turn influence the apparent acidity of H_2CO_3 and solubility of CO_2.

3.1.2.3. The Alkalinity

Sea water can be regarded as the result of the neutralization of acids from volcanoes and bases from weathering action (Sillen, 1961).

Volcanoes effectively liberate important quantities of HCl, H_2SO_4, and CO_2. As to weathering changes, they become expressed chemically as hydrolysis of the silicates, alumino-silicates, and eventually carbonates which release cations in solution, Na^+, K^+, Mg^{++}, Ca^{++}. Opposing these reactions, which may be written schematically:

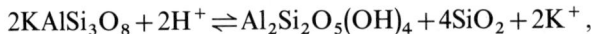

$$2KAlSi_3O_8 + 2H^+ \rightleftharpoons Al_2Si_2O_5(OH)_4 + 4SiO_2 + 2K^+,$$

are the reactions in domains of sedimentation, called aggradations, like:

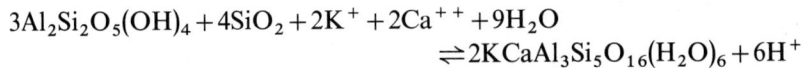

$$3Al_2Si_2O_5(OH)_4 + 4SiO_2 + 2K^+ + 2Ca^{++} + 9H_2O$$
$$\rightleftharpoons 2KCaAl_3Si_5O_{16}(H_2O)_6 + 6H^+$$

that reduces the alkalinity.

The alkalinity is weakly negative in rain water ($SO_4^=$, Cl^-). On contact with rocks, the water acquires a positive alkaline reserve of the order of 10^{-5} to 10^{-3} F kg^{-1} (F = Faradays), the moderating process at higher levels generally limiting the value of R_B to several millifaradays per kg.

3.1.2.4. pH of Several Natural Environments

Rain water charged with various solubles like SO_2, CO_2, and highly hydrolyzable chlorides of heavy metals has a negative alkalinity, and its CO_2 pressure is close to that of the atmosphere. As a consequence, rainwater has a pH between 4 and 5.

Flowing water, and rivers in more sparsely vegetated regions (mountainous areas, for example) have an alkaline reserve ranging from 10^{-4} to 10^{-3} F kg^{-1}. These waters are in equilibrium with atmospheric CO_2 and their pH varies between 7.2 and 8.2.

The alkalinity of sea water is relatively elevated, $R_B = 2.34 \times 10^{-3}$ F kg^{-1} for a chlorinity of 1.9%. In view of the high degree of ionization of the solution, the dissociation constants of H_2CO_3 and HCO_3^- are raised. The pH is weaker than that given by Figure 2-15 and varies only slightly around 8.1.

Water in the soil is relatively acid for two reasons. On the one hand, organic anions corresponding to strong acids diminish the alkaline reserve. On the other hand, the pressure of CO_2 is much elevated. But this low pH increases the speed of weathering of rocks and hence augments the alkaline reserve. As a result a compromise state is reached, with pH only slightly off neutrality and an important quantity of cations in solution.

3.1.2.5. *Influence of pH on the Mobility of the Elements*

Among the most abundant elements of the Earth's crust certain ones like Al and Fe are almost totally absent in water. This can be attributed to the very weak solubility of the hydroxides of aluminum and iron in trivalent form.

Figure 3-3 shows that the solubility of Al is of importance only at extreme pH values rarely encountered in nature. On the other hand, silica has a relatively elevated solubility and small but not negligible amounts can migrate.

3.1.2.6. *Conclusions*

This overall view of acid-base reactions in natural waters may be summarized by the scheme shown in Figure 3-4.

The minor elements sensitive to pH are those that form complexes or an insoluble compound with hydroxide ions or all other anions with basic properties (CO_3^{2-}, phosphates, etc.).

The feedback effect of pH on the carbonate system is important only if the associated gaseous phase is neither infinitely large (atmosphere, $p(CO_2)$ = constant) nor infinitely small (ΣCO_2 constant).

The major ions are affected by the fact that a reduction of pH sets into motion increasing weathering reactions while an increase of pH limits them or replaces them

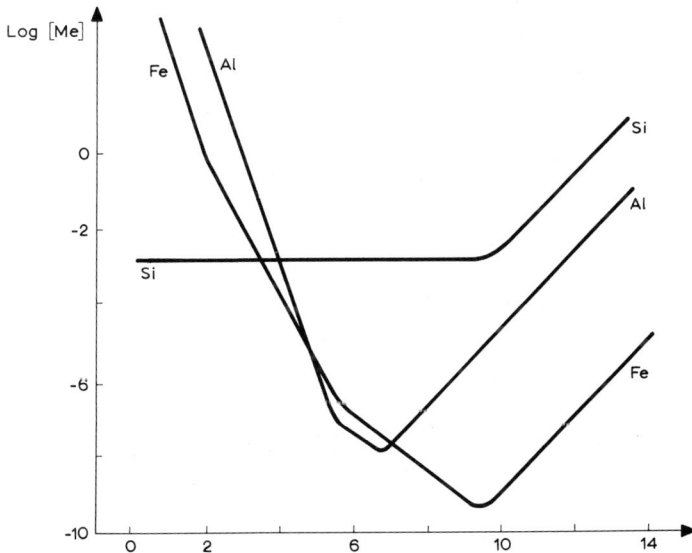

Fig. 3.3. Solubility of silica, alumina, and ferric hydroxide as a function of pH.

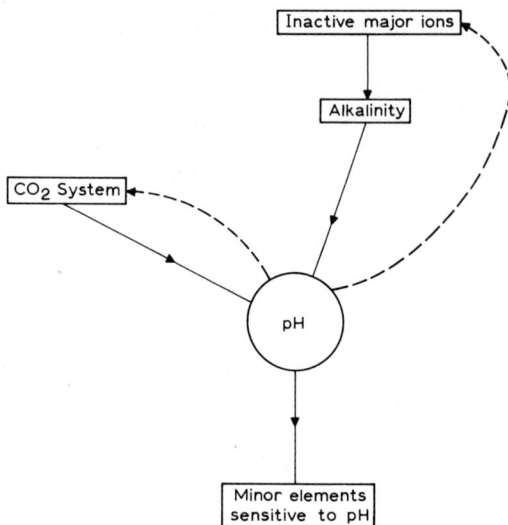

Fig. 3.4.   Pattern of pH.

by aggradation reactions. The last process brings a solid phase into play, and at least in water reservoirs of great size, the speed of response to changes in pH is very sluggish, actually tied to the period of the water cycle (see Chapter 5).

3.1.3. OXIDATION-REDUCTION REACTIONS IN NATURAL WATER

3.1.3.1. *Chief Redox Pairs in Natural Waters*

Among the major ions in natural waters, Ca^{2+}, Mg^{2+}, K^+, and Na^+ are ions of metals that are too highly reduced to exist in water. Cl^- is the ion of a substance that is too strong an oxidant. All of these ions are accordingly inactive in water.

As to CO_3^{2-}/HCO_3^- and SO_4^{2-}, they represent oxidized forms of redox pairs of which the reduced form exists in certain natural conditions that will be described.

Matter that is classifiable as of oxidation-reduction character consists partly of the ions present in trace amounts ($\sim 10^{-8}$ M kg^{-1}) like Fe and Mn, and of partly dissolved matter of which the chief is oxygen. Oxygen is slightly soluble in water, but the atmosphere constitutes an almost infinite reservoir. The mixing of waters and the processes of diffusion virtually assure a homogeneous distribution in the oceans.

In certain tightly closed basins, however, oxygen disappears completely at greater depths. One then finds considerable quantities of H_2S. The transition from the oxygenated zone to the sulfurated zone usually occurs quite abruptly (see the example from the Black Sea in Figure 3-5).

Two distinct cases need to be considered, aerated environments, and anoxic or reducing environments.

3.1.3.2. *Aerated Environments*

Surface waters contain dissolved oxygen essentially in equilibrium with atmospheric

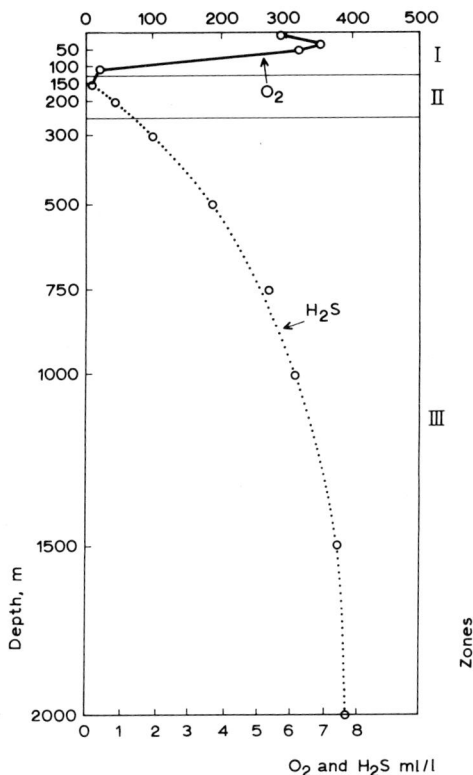

Fig. 3.5. Concentration of gases dissolved in the water of the Black Sea.

oxygen at 10^{-4} M kg^{-1} at 25 °C. The oxidation-reduction (redox) pair – oxygen $(p = \frac{1}{5}$ atm) water $(=1)$ – has a pE given by

$$pE = 20.75 - 0.25 \, pO_2 - pH \simeq 20.6 - pH.$$

Measurements with the platinum electrode taken in the same surroundings give experimental points that scatter rather severely around a value

$$\text{'pE'} \simeq 14 - pH.$$

It is now known that these potentials, measured in this way, are what electrochemists call mixed potentials, to which Nernst's equation is not applicable. Insofar as the theoretical pE alone is meaningful, it predicts whether such an oxidation reaction can or cannot go in water, but it is still not possible to forecast whether an allowed reaction will go or not.

The preceding discussion seems to show that the 'measure' of potential is of no interest. That is not wholly true. On the one hand, some water may contain a rapid redox pair, like Fe^{3+}/Fe^{2+}, in significant amounts; this is the case for water that washes over rocks rich in pyrites. This strongly acid water may contain up to 10^{-2} M kg^{-1} of Fe. The 'pE' measured permits reading the ratio Fe^{3+}/Fe^{2+} in the solution, but in general not the state of oxidation of other pairs present. The solution simply hasn't reached redox equilibrium.

On the other hand, it is amazing that all oxidation reactions that go with a reaction potential higher than that measured go only with the greatest difficulty in natural water. Thus in sea water, Mn can stay in the manganous state, chromium in the trivalent state of oxidation, iodine in the iodide state. But at least locally, oxidation reactions can occur, and reduced states exist only from new influxes. One can thus lay down as a useful, empirical, working rule that oxidation reactions that occur at a potential between the measured value of pE and the theoretical value of pE go very slowly in natural waters.

3.1.3.3. *Reducing Environments*

Reducing environments develop whenever the quantity of organic material is sufficient to consume all the oxygen. Anerobic organisms survive there, reducing sulfate, nitrogen, and carbonates to S, NH_4, and CH_4. All of these pairs that have very low normal potentials (~ -0.3 V at pH 8) are very slow. The measured potential is higher and ranges around 0.1 to -0.1 V. There again, the measured potential has no thermodynamic significance. It was thought that certain pairs, in particular S_n^{2-}/S^{2-} might impose their potential.

In aerated environments to the contrary, it is inordinately difficult to define a theoretical potential because numerous pairs occur simultaneously (SO_4^{2-}/S^{2-}, S_n^{2-}/S^{2-}, N_2/NH_4^+, CO_2/CH_4) and are not ordinarily in equilibrium with one another.

3.1.3.4. *Overall Picture of Oxidation-Reduction Phenomena at the Earth's Surface*

The deviations from equilibrium envisaged above must be considered to be a highly important characteristic of the surface of the planet. The conditions for oxidation and reduction at the surface are the result of the presence of free oxygen in excess and of the existence of organic matter formed by chlorophyll synthesis. This organic material is thermodynamically unstable and tends, whether by other biological processes or by inorganic reactions, toward a state of equilibrium that is in gross a mixture of CO_2 and CH_4. Corresponding to the quantity of organic matter that evolves in this way is a consumption of oxidant. One can associate a capacity for reduction with this quantity of organic matter that can be expressed in Faradays per kilogram of solution.

Insofar as the reducing capacity is less than the available oxygen, the 'thermodynamic' potential of the system is fixed by the pair $\frac{1}{2}O_2/H_2O$. As soon as the reducing capacity exceeds the quantity of oxygen, another oxidant has to be used. Since this oxidant is very weak (whether it is SO_4^{2-}, N_2, or HCO_3^-) the environment becomes strongly reduced.

As is shown in Figure 3-6 there is no intermediate medium in natural waters. Even if the equality between the quantity of oxygen and the reducing capacity is realized at about 10^{-4}, the theoretical potential moves from 0.740 V to -0.180 V. In fresh

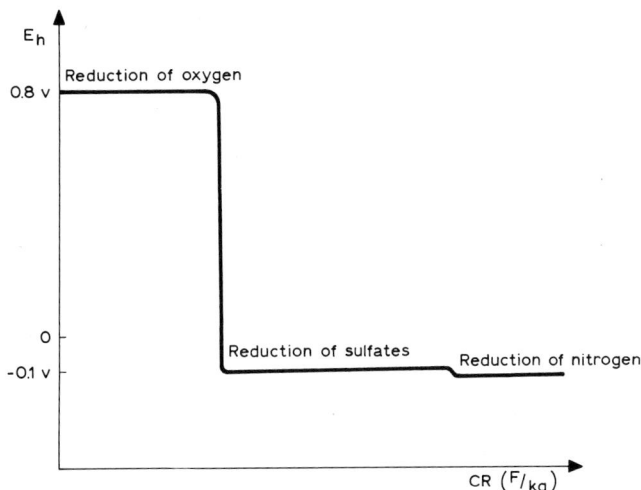

Fig. 3.6. Evolution of the theoretical pE as a function of the reducing capacity of the medium.

sediments, it is possible to have intermediate surroundings due to the reduction by the organic material of solid phases like MnO_2 and FeOOH.

These considerations lead to a picture of the redox properties of natural water that differs from the classical one of the pE–pH diagram due to Baas Beecking *et al.* Instead of reporting the potentials measured in the water as these authors have done (Figure 3-7) we have given the theoretical potentials.

3.1.3.5. *Influence of the State of Oxidation-Reduction on the Mobility of the Elements*

The influence is very great because of the interplay of two independent factors:

(1) For all those elements that can exist in several different states of oxidation (the transition elements) the mobility is large for the highly oxidized forms (CrO_4^{2-}), very weak for intermediate forms (3- and 4-valent) (MnO_2, Fe_2O_3), and fairly large for divalent forms (Fe^{2+}, Mn^{2+}...).

(2) In reducing media, the reduction of sulfates leads to the appearance of important amounts of sulfides that immobilize a large number of metals through formation of highly insoluble salts.

Since the measurements of pE have no thermodynamic significance, it becomes difficult to forecast precisely how the elements sensitive to redox phenomena will behave. Nonetheless, in a certain number of special cases, the precipitation or separation of elements through oxidation-reduction reactions can be explained. A discussion will appear farther on (Chapter 5) concerning the separation of Fe–Mn. As for the precipitation of sulfides, one may imagine attacking the problem by characterizing the natural environment not by its value of pE but by its sulfide content (specific electrodes for S^{2-} ions) and by using solubility diagrams in place of pE, pH diagrams.

3.1.4. FORMATION OF COMPLEXES

The major ions in waters do not form complexes strongly and as a consequence,

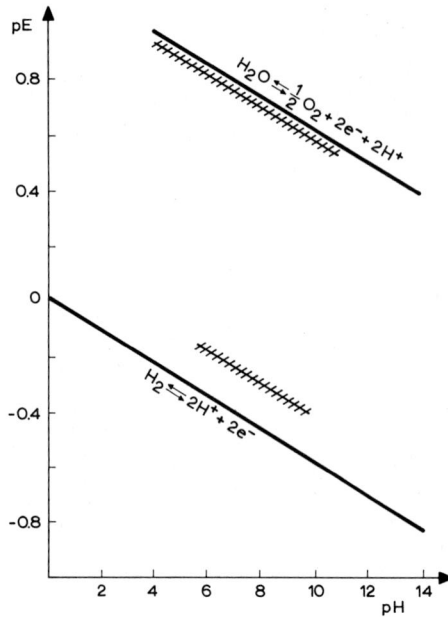

Fig. 3.7.   Diagram of pE vs pH in natural water. The hatched area are the natural systems.

complexes are not very abundant in natural waters. They are surely more important in the most concentrated waters (sea water, brine).

Those that play the most important roles are furthermore not real complexes in the structural sense but associations of pairs of ions that exist only in concentrated solutions that can still be treated thermodynamically as complexes.

3.1.4.1. Combinations of the Major Ions in Sea Water

Sea water contains 8 ions that form more than 99.7% of the mass of dissolved salts (Table 3-II).

Among these ions certain ones can form ion pairs, in particular, $MgCO_3^\circ$, $CaCO_3^\circ$, $MgSO_4^\circ$, $CaSO_4^\circ$, $CaHCO_3^+$, $NaCO_3^-$, $MgHCO_3^+$, KSO_4^-, and $NaSO_4^-$ (in decreasing order of stability). Taking account of these complexes and of activity coefficients measured in the laboratory, Garrels and Thompson (1952) have succeeded in showing that in sea water, 69% of HCO_3^- ions, 54% of sulfate ions, and only 9% of carbonate ions are in free form. For the cations, the formation of complexes is much weaker,

TABLE 3-II

Mean composition in mole l^{-1} of sea water
of salinity 3.45%

Na^+	0.48	Cl^-	0.56
Mg^+	0.054	SO_4^{2-}	0.028
Ca^{2+}	0.010	HCO_3^-	0.0024
K^+	0.010	CO_3^{2-}	0.00027

for 87% of magnesium, more than 90% of calcium, and all of the sodium and potassium ions are in free form. The importance of the complex $MgCO_3^\circ$ must be emphasized as it gives the ion Mg^{2+} a control over the precipitation of calcium carbonate.

3.1.4.2. *Formation of Complexes Among Minor Ion Species*

The activity of the major ions in sea water is relatively constant. The relative importance of complex formation among the minor ions may be judged from a simple table due to Sillen (1961) (Table 3-III).

This table shows the importance of the complex chlorides of cadmium, lead, and mercury that occur in sea water in the forms $CdCl^+$, $PbCl^+$, $HgCl_4^{2-}$, of the fluoride complexes of thorium, the hydroxyl complex of iron $Fe(OH)^{2+}$, or aluminum $Al(OH)_4^-$, of bismuth, of carbonate complexes of uranium $UO_2(CO_3)_3^{4-}$, and of copper $Cu(CO_3)_3^{4-}$.

In ground water, organic complexes assume a non-negligible importance (for example, oxalic complexes) but the data regarding the concentrations of soluble organic material and the constants of complexes are both inadequate.

3.2. Study of the Saturation of Sea Water with Calcite

The two varieties of calcium carbonate, calcite and aragonite, constitute the principal chemical sediments found on present day ocean bottoms (with or without the intervention of biological processes). The map in Figure 3-8a shows, for example, the magnitude of carbonate sediments. A sharp dependence further appears between the percentage of carbonates in the deposits and the depth. The average content is more

TABLE 3-III

pK for complexes in environments with ion strength comparable with that of sea water

Ligand	$-H^+$	F^-	Cl^-	Br^-	$SO_4^=$	$CO_3^=$
log (conc. in sea water)	8.1	-4.2	-0.3	-3.1	-1.6	-3.7
Be^{2+}	-6[a]	5.0[a]	1		0.7	
Sc^{3+}	-5	6.2	1.1	1.2		
La^{3+}	-10	2.7	0		1.5[b]	
Th^{4+}	-4[a]	7.5[a]	0.3		3.3[b]	
UO_2^{2+}	-6[a]	4.5[b]	0		1.8	5.6[a]
Fe^{3+}	-3[a]	5.2[b]	0.6[b]		2.3[b]	
Cu^{2+}	-8[b]	0.7	0		1[b]	5[a]
Zn^{2+}	-9	0.7	0[b]		1	
Cd^{2+}	-10	0.5	1.5[a]	1.6	1[b]	
Hg^{2+}	-3.5[a]	1.0	6.7[a]	9.0[a]	1.3	
Al^{3+}	-5[a]	6.1[b]				
Pb^{2+}	-8[b]		1.0[a]	1.3	1.3[b]	
Bi^{3+}	-2[a]		2.2[b]	2.3		

[a] Complexes highly stable in sea water.
[b] Complexes able to exist in sea water.

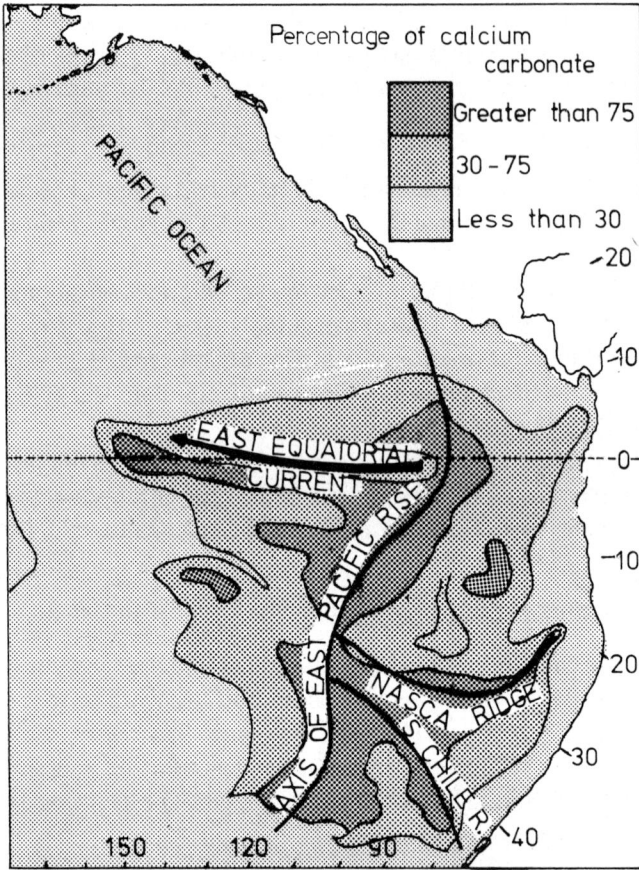

Fig. 3.8a.   Relation between depth and carbonate content of sediments in the Eastern Pacific ($-4-$ = depth of 4 km). [After Bramlette.]

than 80% at depths less than 3500 m and extremely weak (less than 5%) at depths greater than 5000 m, and the transition is most abrupt (Figure 3-8b). The depth boundary varies with geographical locality. It lies deeper in the North Atlantic and shallower in the North Pacific. The problems posed by the deposits of the calcium carbonates and what determines their solubility are highly important. Thus, just as a rough analysis shows that the solubility products are greatly exceeded in the oceans, a finer analysis shows that a large number of factors play important roles such as the ion concentration in the vicinity, the temperature, and the pressure.

3.2.1. ACTIVITY COEFFICIENTS IN SEA WATER

It must first be pointed out that the activity coefficient to be used is not the one usually defined by chemists. Effectively, the precise nature of ions or molecules that contain the chemical element Ca, for example, is often poorly known. One might identify the total amount $(\Sigma Ca = |Ca^{2+}| + |CaSO_4^o| + |CaHCO_3^+| + \cdots)$ by chemical analysis and in some cases the activity of (Ca^{2+}) by measurement of solubility prod-

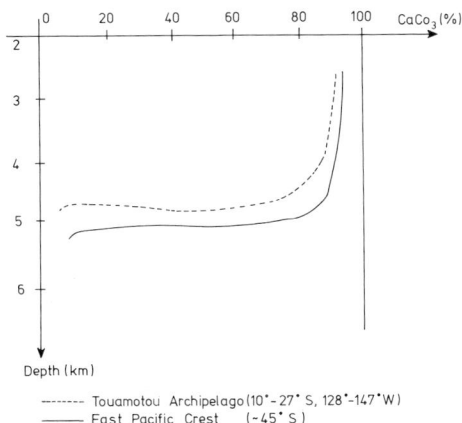

Fig. 3.8b.   Saturation of ocean water with $CaCO_3$ vs depth.

ucts or measurements with a specific electrode. From this information, the geo-chemist most often defines a global activity coefficient by the equation

$$\Gamma_A = \frac{(A)}{\Sigma A}.$$

In sea water, it has proved possible to measure $\Gamma_{HCO_3^-}$ and $\Gamma_{CO_3^{2-}}$ either by precise titration of the sea water with dilute HCl adjusted to the same degree of ionization as the sea water by addition of NaCl or by an elegant method due to Berner (1965). A sample of sea water defined by its carbonate alkalinity A_c (which represents the alkalinity corrected for the small contribution from the borates):

$$A_c - [HCO_3^-] + 2[CO_3^{2-}]$$

is equilibrated in succession with two atmospheres in which the partial pressure of CO_2 is very different and is precisely known. The measured value of pH obtained at equilibrium in each case is inserted in the expression deduced from the preceding one:

$$A_c = \frac{pCO_2}{[H^+]} \left[\frac{\alpha K_{A_1}}{\Gamma_{HCO_3^-}} + \frac{\alpha K_{A_1} K_{A_2}}{\Gamma_{CO_3^-}[H^+]} \right]$$

where α, K_{A_1} and K_{A_2} are respectively the constants in Henry's law relative to CO_2, and the acidity constants of 'H_2CO_3' permit evaluation of $\Gamma_{HCO_3^-}$ and $\Gamma_{CO_3^-}$.

Since sea water is virtually saturated with calcium carbonate, it is possible to add solid $CaCO_3$ to a sample of sea water without altering its composition. The value of pCO_2, of pH, and of ΣCa permits calculation of $\Gamma_{Ca^{2+}}$ if $\Gamma_{CO_3^-}$ has been determined. The results are the following:

$$\Gamma_{CO_3^-} = 0.024 \pm 0.004;$$
$$\Gamma_{Ca^{2+}} = 0.22 \pm 0.02;$$
$$\Gamma_{HCO_3^-} = 0.561 \pm 0.006.$$

These values are in reasonably good agreement with values calculated by Garrels and Thomson by use of Debye-Hückel coefficients and calculating the free fraction of each ion yet keeping in mind the pairs of ions like $MgHCO_3^+$, $CaHCO_3^+$, $NaCO_3^-$, $MgCO_3^\circ$, $CaCO_3^\circ$, $MgSO_4^\circ$, $NaSO_4^- \cdots$.

In another sense, these activity coefficients permit reassessment of the apparent dissociation constant of carbonic acid and solubility constant of the calcium carbonates. These last were determined by Buch and later by Lyman:

$$\alpha' = 2.93 \times 10^{-2}$$
$$pK'_{A_1} = 6.00 \qquad \text{at 25°C for water of chlorinity 1.9\%.}$$
$$pK'_{A_2} = 9.10$$

It is difficult to apply the results of McIntyre and of Zen on the weak dependence of the γ's on temperature and pressure to the Γ coefficients. As a consequence, we will use the apparent constants α', pK'_{A_1}, pK'_{A_2}, etc., and we shall be interested in their variation with salinity, temperature, and pressure (S, T, P).

3.2.2. VARIATION OF THE APPARENT CONSTANTS AS A FUNCTION OF S, T, P

The solubility of calcite and aragonite does not depend solely on the variations of the solubility products of the substances. Indeed, if one considers each element of volume of the ocean as if it were a closed system, the only conservative quantities are ΣCO_2 and R_B, the alkalinity. The pH and the concentration of CO_3^{2-} effectively present undergo variations tied to those of the acidity constant and the carbonic acid constant.

By contrast with what is done in the laboratory where one studies primarily the variations of equilibrium constant with temperature, the variations of temperature in natural water are rather small (with a few exceptions, they are confined to a range between 0 and 30 °C), while the pressure in the ocean may take on considerable values (up to 1 kbar) at great oceanic depths (where an increase of depth of 1 m increases the pressure 0.1 bar).

The variations of α, the solubility coefficient of CO_2 in water, and the apparent variations of K'_{A_1} and K'_{A_2} of carbonic acid, of the apparent K's of the calcium carbonates as a function of the salinity, all these studies have been the subject of numerous experimental studies. The variations of these quantities with temperature are equally well known (Figure 3-9).

The variations with pressure were determined from the dilations ΔV associated with the reaction and further verified experimentally (Distèche and Distèche). The chief results are illustrated in Figure 3-5.

3.2.3. EXPERIMENTAL STUDIES OF THE 'CARBONATE SYSTEM' OF THE OCEAN

In order to determine pH, ΣCO_2, and pCO_2 at the temperature T and total pressure P of the atmosphere, the alkalinity of the carbonates A_c and the alkalinity R_B lead to three relationships:

$$(H^+) = \frac{\alpha pCO_2 K'_{A_1}}{\Sigma CO_2} \left[1 + \frac{(H^+)}{K'_{A_1}} + \frac{K_2}{(H^+)} \right] \qquad (1)$$

$$A_c = \frac{\Sigma CO_2 \left[1 + \dfrac{2K'_{A_2}}{(H^+)}\right]}{1 + \dfrac{(H^+)}{K'_{A_1}} + \dfrac{K'_{A_2}}{(H^+)}} \tag{2}$$

$$R_B = A_c + K_B \frac{\Sigma B}{K_B + (H^+)} \tag{3}$$

ΣB denotes the total amount of dissolved boron ($\Sigma B = 2.2 \times 10^{-5}$) and K_B is the apparent dissociation constant of boric acid. This term is in general very weak compared with the alkalinity but cannot be neglected in a precise study.

It thus becomes necessary to measure two quantities. The ones that can be measured with the greatest precision are ΣCO_2, pCO_2, and R_B, for the variations of pH in this relatively compressed zone are small. The measurements of ΣCO_2 and pCO_2

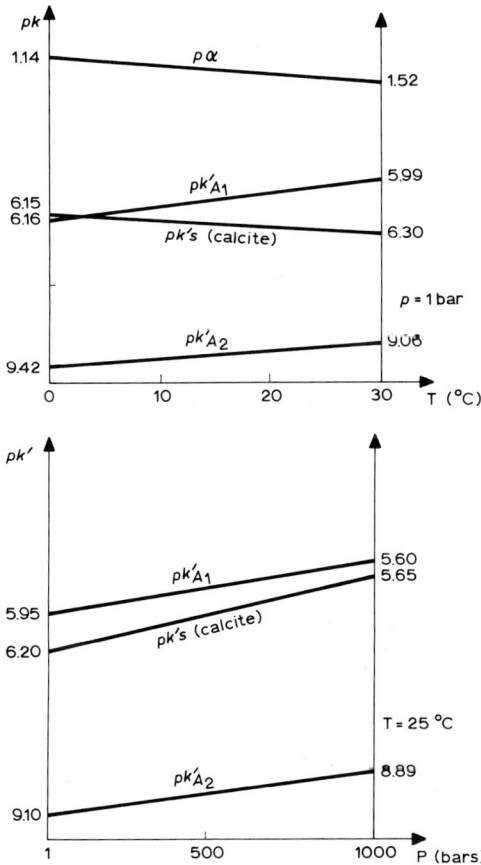

Fig. 3.9. Variation of apparent values of pK of the carbonate system in sea water with temperature and pressure.

can be made with the same apparatus; these are the quantities generally measured. The study by Li *et al.* (1969) serves in particular to illustrate the point.

Through Equations (1), (2), and (3), it has been possible to calculate the alkalinity, which shows relatively little variation. Knowing the content of calcium, one can simultaneously evaluate the ionization product $[Ca^{2+}]$ $[CO_3^{2-}]$ and compare it with the solubility product found in the same conditions of salinity, temperature, and pressure (Figure 3-10).

In the superficial zone, at low latitudes, the temperature is relatively high and the water is supersaturated with calcite. The temperature decreases rapidly between 100 and 1000 m depth, thereby leading to a decrease in the degree of saturation. The de-

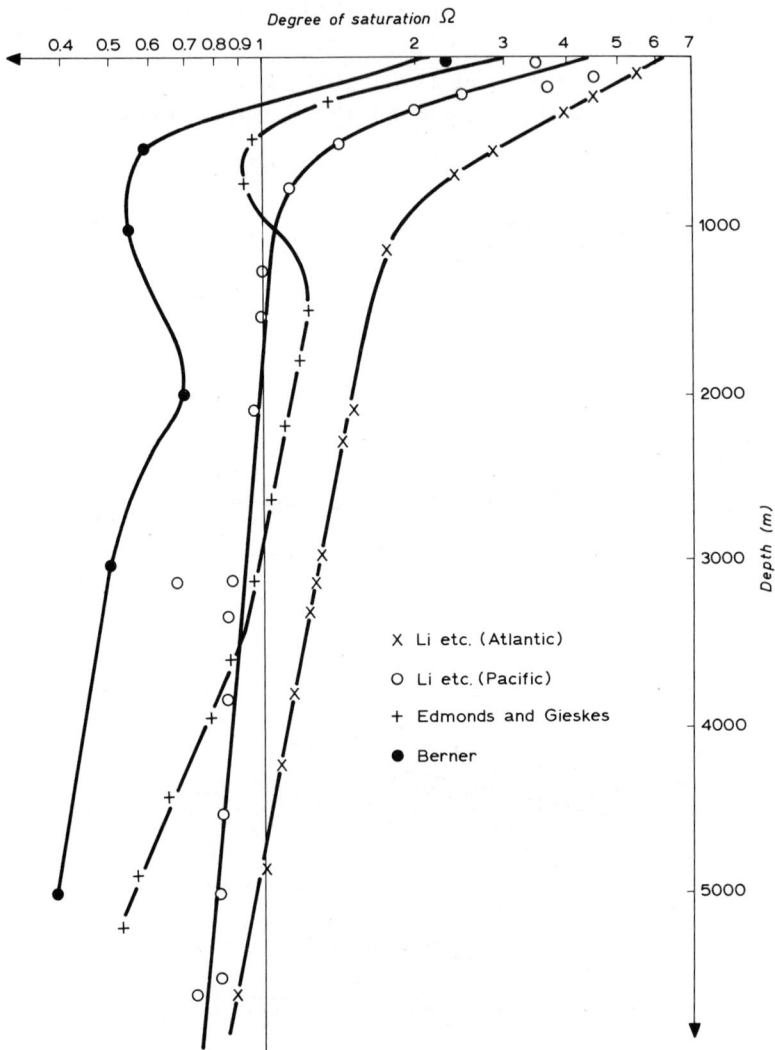

Fig. 3.10.   Degree of saturation of the oceans with calcium carbonate.

crease is reinforced by the production of important CO_2 in the vicinity of 1000 m. The slow decrease at greater depths is due to the effect of pressure.

Pacific Ocean waters are saturated with aragonite at depths less than 200 m and with calcite at depths less than 2300 m. The corresponding Atlantic depths are 2000 m and 4500 m. Measurements by Peterson (1961) confirm these results. On suspending crystals of very pure calcite in Pacific waters, they show no change whatsoever at depths less than 3000 m. At greater depths than 3700 m they dissolve very rapidly.

It is in any case necessary to note that the compensation level is at a lesser depth than that corresponding to the rapid dissolving of the calcite. This difference, which is tied to the kinetics of dissolving of the calcite, becomes, in fact, the objective of a precise study.

The explanation thus emerges of the fact we cited at the start of this study that the sediments at greater depths than 4000 m contain no $CaCO_3$.

3.2.4. CALCITE OR ARAGONITE

The preceding study shows in particular that surface waters are supersaturated by a factor of 6 with calcite and by a factor of 3 or 4 with aragonite. In addition, at all points in the ocean, calcite is less soluble – hence more stable – than aragonite. Now aragonite is often found in actual marine sediments. These sediments contain a non-negligible component of debris of sea shells that may be composed either of pure calcite (oysters) or pure aragonite (coral), or still further of a mixture of the two and even a small amount of a third variety of $CaCO_3$: vaterite. The directly precipitated carbonate is also often aragonite.

Numerous laboratory tests have brought out the influence of magnesium ions. Attempts to precipitate $CaCO_3$ starting from an aqueous solution containing very small concentrations of magnesium ions ($\sim 10^{-3}$ M) led to precipitates containing more than 75% aragonite. The ion Mg^{2+} inhibits the nucleation of calcite (Pytkowicz). Studies by Kitano et al. (1965) also show the influence of organic compounds on the nature of the carbonates that are formed. The compounds forming complexes with calcium (e.g., pyruvates, citrates) favor the precipitation of calcite. Others, by unknown mechanisms, favor the precipitation of one or both species. Still others (acetates, dextroses) have no influence. Many of the compounds studied by Kitano et al. (1965) play a role in animal metabolism and lead to variation in the composition of shells.

Aragonite once formed can last for very long times. But in ancient sediments, most of the aragonite formations (whether shells or not) have recrystallized into calcite.

3.2.5. BALANCE OF CALCIUM IN THE OCEAN (Li et al., 1969)

Estimates by Livingstone lead to the transport of 10^{13} moles of Ca yr^{-1} by streams and rivers into the ocean. The quantity of $CaCO_3$ redissolving in the depths of the oceans has been estimated by Li et al. (1969) as $(6 \pm 2) \times 10^{13}$ mole yr^{-1} depending on differences between the alkalinity at the surface and in the deep zone.

If it is accepted that the quantity of Ca in sea water is in a stationary state, it follows that the total precipitation of calcium in higher-level waters is $(7 \pm 2) \times 10^{13}$ mole yr^{-1} of which 15% deposits and the rest redissolves.

We can compare this figure of 15% with that at the surface of the ocean bottom that is bathed in unsaturated $CaCO_3$, and with those bathed in supersaturated $CaCO_3$. Approximately 80% of ocean bottoms lie below the compensation depth for calcite and almost 100% below that for aragonite.

Finally, it is the ratio between the quantity of calcium added by weathering and the quantity fixed by organisms in superficial zones of the ocean that fix the global level of compensation. The spatial variations of the compensation depth are controlled by the production of CO_2 from organic carbon; this in turn is more important in the Pacific than the Atlantic, the level occurring at lesser depths in the former than in the latter.

References

Berner, R. A.: 1965, *Geochim. Cosmochim. Acta* **29**, 947.
Edmond, J. M. and Gieskes, J. M. T. M.: 1970, *Geochim. Cosmochim. Acta* **34**, 1261.
Garrels, R. M. and Thompson, M. E.: 1952, *Am. J. Sci.* **260**, 57.
Li, Y. H., Takahashi, T., and Broecker, W. S.: 1969, *J. Geophys. Res.* **74**, 5507.
Michard, G.: 1967, *Mineral Deposita* **2**, 34.
Morris, J. J. and Stumm, W.: 1967, *Equilibrium Concepts in Natural Water Systems*, p. 270.
Sillen, L. G.: 1961, *Am. Assoc. Advan. Sci. Pub.* **67**, 549.

CHAPTER 4

THE GEOCHEMICAL FRACTIONATION OF
TRACE ELEMENTS

Introduction

Just as the major elements do, the trace elements also suffer dispersion or fractionation in the course of geochemical processes. They become distributed among phases of which they are not the essential constituents and which they did not create. They can take part in catalyses but this role, which may be quite important, is still totally unknown.

The dispersions of the traces are among various solid and liquid phases where they are in solution. These fractions obey relatively simple laws which are embodied into the theoretical models for geochemical processes. The models, constructed from petrographic observations, are refined gradually as observations and laboratory experiments accumulate. In this way our knowledge of the distribution of the elements progresses step by step, and in fact of geochemical processes themselves. The trace elements then serve in their turn as tracers for the geochemical phenomena.

Here again, we consider on the one hand the magmatic processes and on the other the behavior of the trace elements in the hydrosphere.

Partition of a Trace Element Between Two Phases

At equilibrium, the chemical potential of the trace element must be the same in both phases

$$\mu_i^\alpha = \mu_i^\beta.$$

Here μ_i^ϕ is the chemical potential of the element i in phase ϕ and may be written

$$\mu_i^\phi = \mu_{0_i}^\phi + RT \log \gamma_i^\phi C_i^\phi$$

where C_i^ϕ is the concentration of element i in phase ϕ, γ_i^ϕ are activity coefficients, and μ_{0i}^ϕ is a constant. After transformation, one obtains:

$$\frac{C_i^\alpha}{C_i^\beta} = \frac{\gamma_i^\beta}{\gamma_i^\alpha} \exp\left[\frac{\mu_{0i}^\beta - \mu_{0i}^\alpha}{RT}\right].$$

If γ_i^β and γ_i^α change little with concentration, one may at a given temperature, regard the right side of the equation as constant and hence define the partition coefficient of the element between phases α and β by

$$\frac{C_i^\alpha}{C_i^\beta} = K_i^{\alpha\beta}.$$

However, the trace element substitutes for a determined major element in the phase. It is therefore preferable, and in fact essential, especially when one of the phases is an aqueous solution, to use a coefficient of exchange between the trace T_i and the major ion M_j, according to the reaction:

$$T_i^\alpha + M_j^\beta \rightleftharpoons T_i^\beta + M_j^\alpha.$$

Then

$$\frac{C_i^\alpha / C_j^\alpha}{C_i^\beta / C_j^\beta} = K_{ij}^{\alpha\beta},$$

with

$$K_{ij} = \frac{\gamma_i^\beta / \gamma_j^\beta}{\gamma_i^\alpha / \gamma_j^\alpha} \exp\left[\frac{\mu_{0_i}^\beta - \mu_{0_i}^\alpha - \mu_{0j}^\beta + \mu_{0j}^\alpha}{RT}\right].$$

This expression can obviously be derived from statistical considerations and the Boltzmann distribution law. One can then see that the μ_0's are the energies of the two types of atoms in the different phases. We shall return to this aspect of the partition coefficients.

4.1. Fractionation of the Trace Elements in Magmatic Processes

4.1.1. MODELS OF THE PARTITION OF THE TRACE ELEMENTS DURING THE MAGMATIC PROCESSES

Among the possible interactions between magmas and solids, the most important are fractional crystallization and partial melting. There are others, too: gaseous transfers, zone melting, exchanges between the magma and reservoirs. These, however, have been far too scantily studied for us to treat them here. We may, however, speak of mixtures between magmas that are important to the relationship between the crust and the mantle.

4.1.1.1. *Model of Fractional Crystallization*

When a mass dm of magma crystallizes, a quantity dx of the trace element enters into the resulting solid. If the crystallization occurs in equilibrium with the magma (surface equilibrium)

$$\frac{dx}{dm} = k\frac{x}{m}$$

x being the amount of trace element remaining in the magma whose mass is m, and k being the partition coefficient of the element between solid and magma. Integration leads to an expression for the concentration C_M of the trace element in the magma

$$C_M = C_0 F^{k-1},$$

where F denotes the fraction of the magma not yet crystallized. The concentration of the element in the solid is

$$C_S = k C_0 F^{k-1}.$$

This expression depicts the evolution of the concentration from the center to the periphery of a mineral crystal that may be formed. The study may be carried out point for point, for example, by means of the electron microprobe.

On the other hand, if a mineral is analyzed, a mean value is obtained that can be expressed as

$$\overline{C_S} = \frac{C_0}{1-F} \int_1^F k\, f^{k-1}\, df = C_0 \frac{1-F^k}{1-F}.$$

This very simple law explains the impoverization or enrichment of an element in one of the two phases. One can see from Figure 4-1 that if F is large enough, one obtains considerable fractionation.

We have assumed so far that k is constant and this is rarely true. The value of k varies first because the temperature varies, then because the nature of the minerals formed changes also.

The variations of k as a function of temperature can also be known but it is also necessary to know how the temperature changes in the course of the crystallization. This in turn brings into play the latent heats of crystallization, the form of the magmatic body, the temperature of the envelope....

A first approximation is to picture a mean temperature of crystallization and to consider the sequence in which the minerals crystallize. When several minerals crystallize simultaneously, the partition coefficients may be generalized in the form:

$$\overline{k} = \lambda_1 k_1 + \lambda_2 k_2 + \cdots + \lambda_y k_y + \cdots + \lambda_n k_n,$$

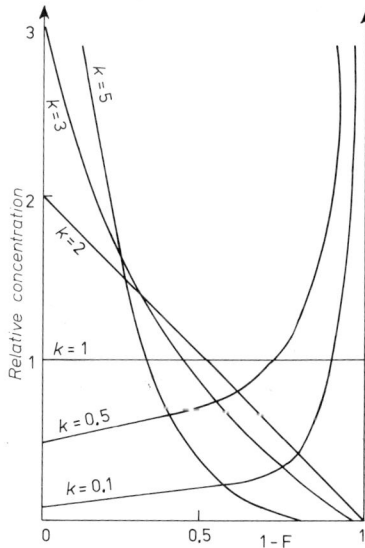

Fig. 4.1. Concentration of an element in the solid phase as a function of (1-F.) (F is the degree of crystallization).

λ_γ designating the fraction of mineral γ in the aggregate and k_γ the partition coefficient between the mineral γ and the magma.

By way of example, the evolution of Rb content may be given (Figure 4-2) in a magma that is precipitating to anorthite, then to anorthite and leucite.

The growth of content of a trace is still more complex when one of the minerals formed in the course of a first step redissolves by reacting with the melt. Thus a basic magma leads first to olivine which then reacts with the molten bath to form pyroxene. Figure 4-3 shows the behavior of nickel in the course of this crystallization. The nickel first is strongly fixed in the olivine; the melt at first becomes impoverished in nickel; the partition coefficient being less favorable for the pyroxene, the transformation of the olivine to pyroxene therefore leads to an increase of nickel in the melt.

4.1.1.2. *Model for Partial Melting*

This time one starts from an assemblage of solid phases of determined composition and investigates the composition of the melt obtained from melting of a fraction of the solid.

Assuming that for the trace element there is equilibrium between the solid and the magma one may write

$$\frac{x_0 - x_m}{S_0 - M} = \bar{k}\,\frac{x_m}{M},$$

where x_0 and x_m are the quantities of the trace element in the initial solid and in the magma and S_0 and M are the total initial masses of solid and magma, $\bar{k}$ is the effective partition coefficient of the process.

If one calls $\phi = M/S_0$ the fraction of the solid that is molten, the concentrations in

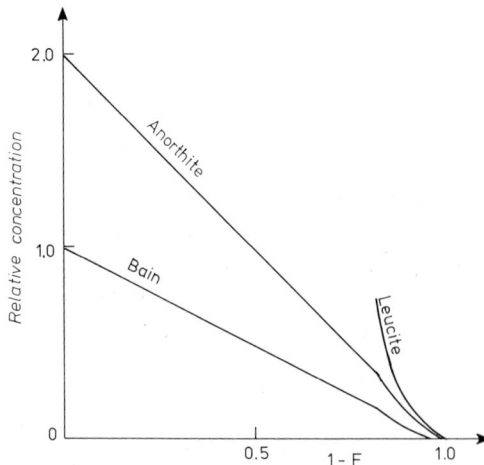

Fig. 4.2.   Evolution of the concentration of a trace element in the course of crystallization from a magma that yields anorthite followed by leucite (from Neuman *et al.*, 1954).

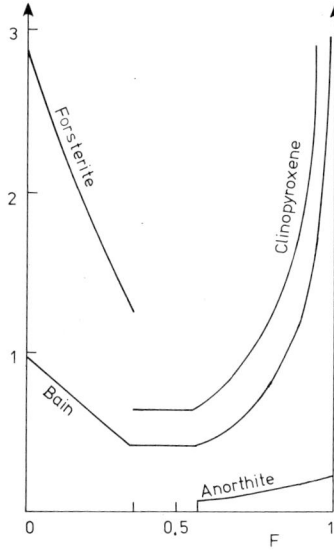

Fig. 4.3. Evolution of the concentration of a trace element in the course of crystallization from a magma that yields olivine, pyroxene, and anorthite (from Neuman *et al.*, 1954).

the magma and the solid are expressed by

$$C_M = \frac{C_0}{\overline{k} + (1 - \overline{k})\,\phi}$$

$$C_S = \frac{\overline{k}C_0}{\overline{k} + (1 - k)\,\phi},$$

The variation of C is a hyperbolic function of ϕ. Figure 4-4 shows that for very small molten fractions, C_M/C_0 can be rather large. It will be desirable to compare the distributions thus obtained with those from fractional crystallization.

Contrary to the preceding case, melting leads generally to a eutectic, the temperature does not vary, and the k_γ's can be regarded as constants. On the other hand, the nature and proportion of the minerals that melt vary and the k coefficient is variable.

One can then consider (Shaw, 1970) that the assemblage of minerals is formed from phases $\alpha, \beta, \ldots, \gamma, \ldots, \lambda$ and in mass fractions $X_0^\alpha, \ldots, X_0^\gamma, \ldots, X_0^\lambda$ with $\sum_\gamma X_0^\gamma = 1$. In each phase, the concentration of the element is C_γ, the composition of the solid initially is thus

$$C_0^S = \sum_\gamma C^\gamma X_0^\gamma.$$

During the fusion, each phase contributes to the formation of the magma in proportion to p^γ, with $\sum_\gamma p^\gamma = 1$.

In regard to the partition coefficients $K^{\alpha\beta}$ among phases α and β and $K^{m\alpha}$ between

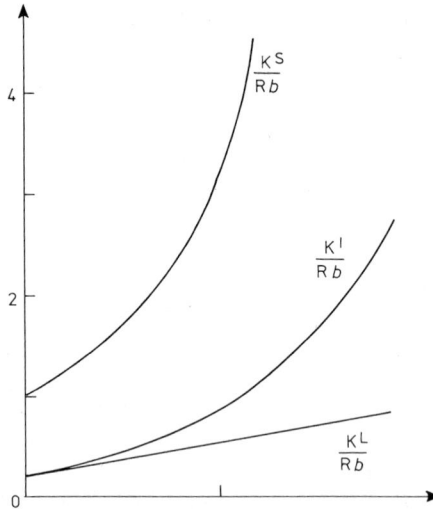

Fig. 4.4. Evolution of the ratio K/Rb during a Rayleigh distillation of a basaltic magma (after Shaw, 1968). 1. Liquid momentarily in equilibrium with the residual solid, then isolated; L, Liquid standing in equilibrium with the residual solid; S, Solid.

magma and phase, the parameters may be defined:

$$P = \frac{p^\alpha}{K^{m\alpha}} + \frac{p^\beta}{K^{m\alpha} K^{\alpha\beta}} + \cdots,$$

$$D_0 = \frac{X_0^\alpha}{K^{m\alpha}} + \frac{X_0^\beta}{K^{m\alpha} K^{\alpha\beta}} + \cdots.$$

One may write:

$$C_M = \frac{C_0}{D_0 + \phi(1 - P)}.$$

This formula is quite similar to the one established earlier. If the initial proportions of the phases and the eutectic composition intervene, the possible hypotheses are limited to the mineralogical composition of the initial solid. The compositions of the eutectic liquids are known from laboratory experiments if the pressure is known at which melting occurs. The partition coefficients can equally well be obtained from laboratory experiments, but as will be shown presently, this is not always easy.

Figure 3-10 shows the evolution of the concentrations of K and Rb as a function of ϕ for different assemblages that can be found in the upper mantle.

It is equally possible to envisage numerous successive partial fusions (the C_0 for the nth fusion being the C_S of the $(n-1)$ st.) Schilling and Winchester (1967) have shown that the concentration in the melt can be put in the form:

$$(C_M)_n = \frac{C_0 \prod_{i=1}^{n} \overline{K}_i}{\prod_{i=1}^{n} y_i + \overline{K}_i(1 - y_i)}.$$

4.1.1.3. *Mixing Model*

If two magmas of masses M_1 and M_2 become mixed, the concentration of the trace element in the mixture is expressed by

$$C_\Sigma = \frac{x_1 + x_2}{M_1 + M_2},$$

where on substituting

$$\mu_1 = \frac{M_1}{M_1 + M_2}$$

$$C_\Sigma = C_1 \mu_1 + C_2 (1 - \mu_1).$$

This case is observed when a basic magma digests a piece of pre-melted crust and assimilates it completely.

To test such a model, one may know or estimate C_1 and C_2 for several elements; upon measuring C_Σ, a value of μ_1 is obtained for each element, and the model is acceptable as correct if the values of μ_1 are the same.

The different models for magmatic evolution can thus be translated into mathematical form. The importance that devolves on the determination of the partition coefficients for knowledge of magmatic phenomena can be seen.

4.1.2. DETERMINATION OF THE PARTITION COEFFICIENTS

There are two ways for determining in practice the partition coefficients of the trace elements under magmatic conditions: by laboratory experiments or by determining values of the coefficients as found in natural assemblages.

(1) In the laboratory, it is attempted to put two minerals into equilibrium, or one mineral and one liquid (magma and water) and bring the content of trace element in the two media into equilibrium. Such experiments are not numerous to date and we shall see the difficulty they present.

The attainment of equilibrium is always debatable in experiments in mineralogical synthesis. All researchers agree that when the reaction of partition stops evolving, equilibrium has been reached, but it is well known that this may in fact be only a metastable equilibrium. On comparing the results with measured natural values, one is also testing the equilibrium, but in every case, the attainment of equilibrium remains open to question. To assure faster kinetics for the phenomena, water is used as it is well known to speed the reactions greatly. Thus Iiyama has put the following technique for measuring the partition coefficient among minerals into use.

Two pods each enclosing a mineral are placed in an enclosure filled with water vapor. To this a radioactive tracer is added (for instance, to study the partition of Rb, ^{89}Rb ...) and the isotopic exchange is utilized. The choice of minerals for M_1 and M_2 is difficult. Whether to use natural minerals, which must then be carefully analyzed, or alternatively as Iiyama (1968) did, use synthetic minerals which do not meet natural conditions and in particular ignore completely the effect of competition of ions for available crystal sites is a problem. In every case, these experiments are delicate and they are only at their start; in particular, the crystal-magma partitions are still incompletely explored by this route (Drake and Weill, 1973).

Nevertheless, there is no doubt that this domain must be explored, and with enthusiasm, during the coming decade.

(2) In natural magmatic rocks, the concentration of diverse elements in diverse selected minerals is measured mechanically. The glass is supposed to constitute the magma.

Schnetzler and Phillpotts (1970) undertook this sort of determination. Their results show that considerable variations are observed in the determinations of partition coefficients made in this way (Table 4-I) but nonetheless the orders of magnitude remain unaltered. The difficulties are correspondingly great in this method.

– Equilibrium is perhaps not always attained particularly in the case of rapidly crystallized rocks where the diffusion of ions in the magma comes into play;

– The purity of the minerals by mechanical and magmatic separation is not always easy to attain;

– Zoning phenomena resulting from the phenomenon of fractional crystallization are met with at the scale of the minerals, and the extraction eliminating the external portions of the minerals can completely falsify the determination of the partition coefficients (Albarède and Bottinga, 1971; Chapter 5).

4.1.3. VARIATIONS OF THE RATIO K/RB IN NATURE

The chemical properties of the major element K and the trace element Rb are closely related. The two elements lie one over the other in the first column of the periodic table. One might accordingly imagine that the behavior of Rb would be a close copy of that of K and that the ratio K/Rb would be constant. This conclusion was in fact reached in early, still highly imprecise studies.

Subsequently, it proved that the ratio could vary between wide limits, from 200 in terrestrial acid rocks and carbonaceous chondrites to 3000 in sea water, with all intermediate values. Since K is a good indicator of rock types, the ratio K/Rb is graphed as a function of K. Figure 4-2 shows the pattern of ratios for terrestrial and

TABLE 4-I

Partition coefficients between minerals and host lavas
(Schnetzler and Philpotts, 1970)

Mineral	Ce	Nd	Sm	Eu	Gd	Dy	Er	Yb
Amphibole:								
Min.	0.18	0.16	0.24	0.26	0.28	0.31	0.24	0.23
Max.	0.26	0.40	0.56	0.63	0.66	0.62	0.47	0.30
Clinopyroxène:								
Min.	0.18	0.26	0.38	0.39	0.46	0.50	0.46	0.43
Max.		0.32	0.43	0.48	0.57	0.56	0.53	0.48
Olivine	0.009	0.007	0.006	0.006		0.009	0.009	
Plagioclase		0.11	0.10	0.35	0.09	0.09		

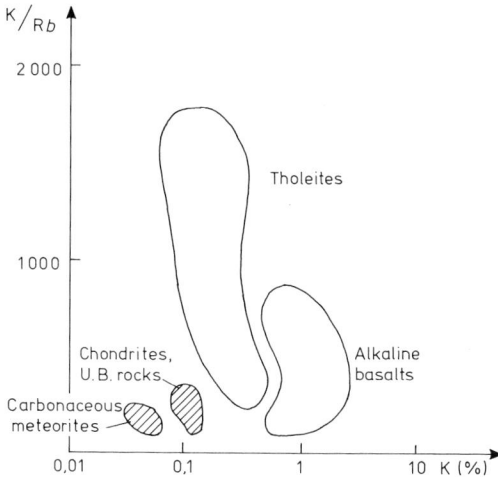

Fig. 4.5. Graph of K/Rb vs K for basalts and meteorites (after Gast (1968) modified).

for extra-terrestrial rocks. One sees at once that ultrabasic rocks and chondrites occupy analoguous positions. The same is true for tholeitic basalts and achondrites, all of which poses the following problems:

(1) What are the relationships among tholeites, alkaline basalts, and ultrabasic terrestrial rocks?

(2) What is the composition of the mantle? Of chondritic types? How is it possible to produce tholeites? Achondrites? How is the formation of ultrabasic rocks to be explained? (What are the relations between chondrites and achondrites? It is a problem that is met at almost every turn but which is outside the scope of this work.)

On the whole, the majority of these questions have not yet received satisfactory answers. Some interesting results have already been recovered but it will require multiplication of the measurements of partition coefficients and detailed studies of the volcanic and plutonic series. The most important results are the following:

(1) Starting from a basic magma where K and Rb are minor elements, the differentiation by fractional crystallization of pyroxenes, olivines, and plagioclases leaves the ratio of K/Rb virtually constant although the content of K varies. This result is obtained equally well starting from partition coefficients and by the study of petrographic series which one can further show were formed by fractional crystallization. (Skaergaard Intrusion, Dolerites of Tasmania).

(2) As previously noted, the only mineral of basic rocks that sharply alters the ratio of K/Rb is hornblende (and not so sharply the plagioclase). Starting from this information, Gast (1968) showed that the transition from tholeites to alkaline basalts could not occur by fractional crystallization. To explain the observations, a 90% crystallization of hornblende would have to occur, and this is regarded as practically impossible.

(3) A model that takes into account the observed facts is one that admits that tholeites result from an important partial fusion (30%) of the ultrabasic mantle while

at the same time alkaline basalts arise from a much more moderate fusion (3 to 5%).

While caution is necessary in assigning significance to values derived from oceanic rocks because of the risk that exchanges with sea water may have occurred (Hart, 1969), Gast (1968) nevertheless believes that the values for tholeites border on those for the upper mantle which would thus be of achondritic type or depleted by several episodes of partial meltings. This view however is disputed by Russel and Ozioma (1971) who believe that the maximum value of the ratio of K/Rb in the mantle is 710.

(4) The difficulties of analysis of ultrabasic rocks make the transition from ultra-basic rock to basalt still obscure. In the special case of the ophiolitic complexes, it has been possible to show how transition from a basic sequence to ultrabasic rocks can occur. The peridotites can be conceived as residues of deep-lying partial fusion. The crystallization would continue at the surface during the extrusion of the complex to give the basic series, the appearance of plagioclases causing the discontinuity in the K/Rb ratio.

It must finally be noted that modern researches with the microprobe show that most of the time, K and Rb are present in cracks between and inclusions in crystals. This observation drastically limits the use of these tracers in basic and ultrabasic rocks with a hypothesis that they are in equilibrium and the use of corresponding partition coefficients.

4.1.4. Differentiation of the rare earths

4.1.4.1. *Diagrams by Coryell et al. (1963)*

In order to bring out geochemical differentiation more clearly, Coryell *et al.* (1963) had the idea to normalize the abundances of the elements in rocks to those in chon-drites. By this means, the greater abundance of elements with even atomic numbers compared with odd ones was eliminated.

For the rare earths, the variation of abundances thus normalized is plotted as a function of atomic number, and the resulting curves are highly regular both for terrestrial and for extraterrestrial rocks (Figure 4-6). The extraterrestial curves are essentially horizontal; there is no differential separation, except for Eu, and we shall return to this element later. In terrestrial rocks, which are generally richer than the chondrites, it is possible to have a rather important relative differentiation, and the behavior of the cerium types is generally different from the yttrium types. This in fact leads to the use of a La/Yb diagram, showing this ratio as a function of the total rare-earth content, as a means of characterizing terrestrial rocks (Figure 4-7).

4.1.4.2. *Europium and Cerium*

The rare earths occur in nature in oxidized form with valence 3. In general, this state is quite stable, and the rare earths do not change their state of oxidation. Two rare earths, however, are exceptions to this rule; they are Eu and Ce. Europium is reducible from Eu^{3+} to Eu^{2+}. When it occurs as Eu^{2+}, it closely resembles Ca^{2+}, both in ionic radius and charge, and as a result tends to follow Ca and separate from the other rare

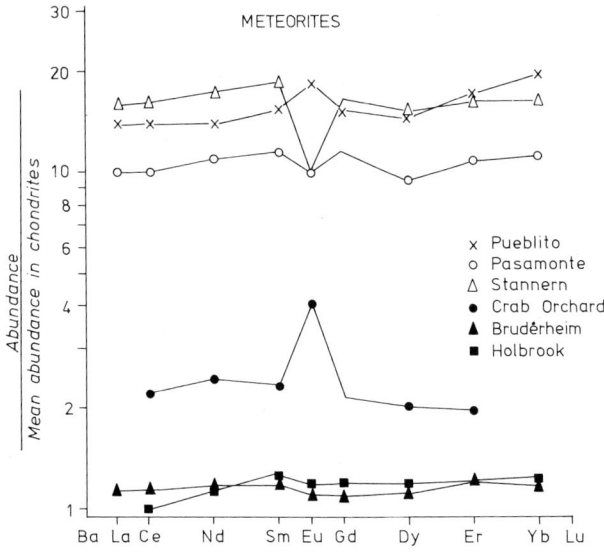

Fig. 4.6. Coryell diagram for several meteorites (after Gast, 1968).

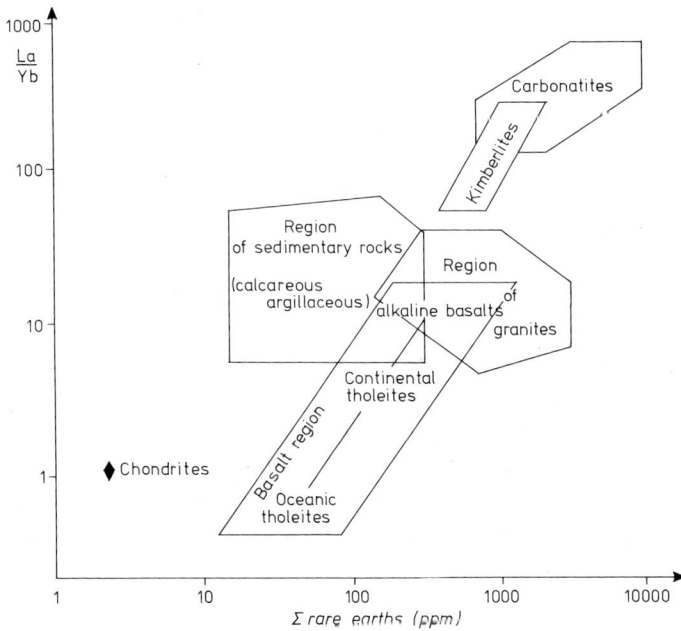

Fig. 4.7. La/Y as a function of total rare earth content in terrestrial rocks.

earths. Thus, for example, Eu tends to concentrate in the plagioclases, in particular the calcic ones. Hence the curves of abundances of rare earths in plagioclases show a positive anomaly for Eu.

Extending this characteristic further, the rocks that result from even a partial accumulation of plagioclase show a positive anomaly in Eu.

Conversely, the rocks that formed after the deposition of plagioclase show a negative anomaly in Eu. All this is true provided that the environment in which these phenomena took place was sufficiently reduced to allow adequate amounts of Eu^{2+} to be present.

The reading of the Eu anomaly can thus serve as an index of the oxidation-reduction state of the environment in which the rocks were formed.

For Ce the phenomenon is inverted. Cerium oxidizes relatively readily to Ce^{4+}. There again, this Ce^{4+} differentiates itself from the other rare earths preferentially, giving a positive or negative Ce-content anomaly. Thus for example, the carbonatites which apparently formed under oxidizing conditions usually show negative Ce anomalies.

4.1.4.3. *The Genesis of Basic Terrestrial Magmas*

As we have already pointed out, there are two principal families of terrestrial basalts: tholeitic (or subalkaline) basalts and alkaline basalts.

To eliminate any interpretive complication that might creep in from possible contamination by the granitic crust, we shall attack the problem posed by this duality with oceanic rocks.

The distribution of rare Earths in tholeites and alkaline basalts is very different (Figure 4-8). The explanation of this difference at the same time is a contribution to the petrogenetic problem posed in Chapter 2. In the guise of hypothesis, one can imagine the alkaline basalts derived from tholeitic basalts by fractional crystallization. The balance of necessary major elements is known (G. D. MacDonald) and the mineral-magma partition coefficients for the rare earths (Figure 4-9). It is possible to exclude: (1) Garnet which is not a mineral that enters into play in such a process; (2) plagioclase, which if it plays any part, creates a negative Eu anomaly in alkaline basalt; (3) olivine which does not noticeably separate the heavy rare earths relative to the light ones. Thus there remain only the pyroxenes as responsible for this process, and in particular, orthopyroxene.

As has been well pointed out by Gast (1968), it is rather difficult to reconcile quantitatively the foregoing positive finding with that arising from the phase diagrams relating to the major elements. If one retains the phenomenon of magmatic differentiation as the means of transition from one type of basalt to the other, it is necessary to bring in another process than simple fractional crystallization (it might, for example, be a gaseous transfer).

One can also, following Gast, turn to another approach and consider that the two types of basalts represent two different degrees of partial fusion of the mantle.

Naturally, the exact chemical and mineralogical composition of the mantle is not

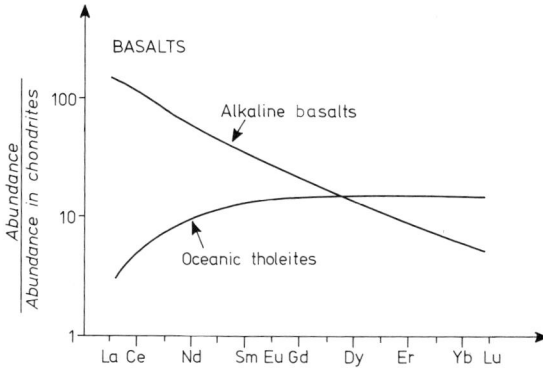

Fig. 4.8. Typical distribution of rare earths in an alkaline basalt and an oceanic tholeite.

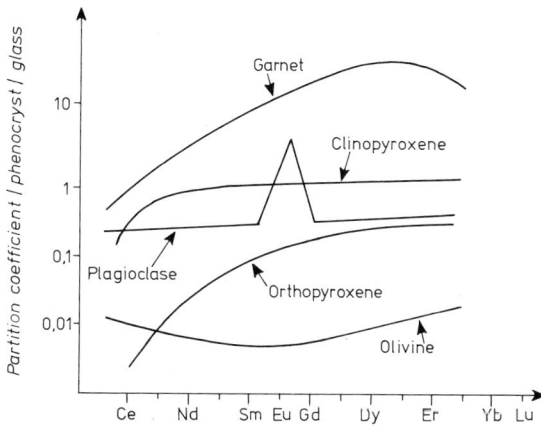

Fig. 4.9. Partition coefficient of minerals/glass for several important minerals.

known. However, we have seen that a reasonable hypothesis can be made as to its constitution (an association of pyroxene + olivine + garnet + spinel, an association called pyrolite).

Starting from such models, Gast, utilizing on the one hand the liquid-crystal partition coefficients, on the other the composition obtained from equilibrium phase diagrams, has shown that the observed facts could be explained, provided:

(1) Tholeites result from 15 to 30% partial fusion of the mantle.

(2) The alkaline basalts only arise from 5% fusion of the same mantle.

Since, incidentally, various authors have established that the degree of partial fusion decreases with pressure, Gast deduced therefrom that alkaline basalts are formed at greater depths than tholeites.

This manner of thinking seems to agree entirely with that defended by Kuno to explain the zoning of island arcs.

It is doubtless too early to make definite conclusions on this subject, but once more, our purpose is above all intended to illustrate a mode of reasoning.

4.1.5. THE TRANSITION ELEMENTS

4.1.5.1. *Use of Coryell et al.'s* (1963) *Diagram*

Using the same normalization procedure of Coryell *et al.* (1963), Allègre *et al.* (1968) studied the distribution of the transition elements in the Earth's crust. The resulting distribution as a function of atomic number, for the majority of terrestrial rocks, looks in graphical form like a large letter W (Figures 4-10 and 4-11).

More precisely, when one progresses from the peridotites to the granites, the W-like shape of the curve grows more and more pronounced.

This evolution is also very marked when one considers certain particular magmatic series, the volcanic series of Hawaii or of Hérault (France) or the pluto-volcanic complex of the Skaergaard Intrusion.

4.1.5.2. *Crystal-Field Theory*

The interpretation of these facts can be approached through the theory of the crystal-

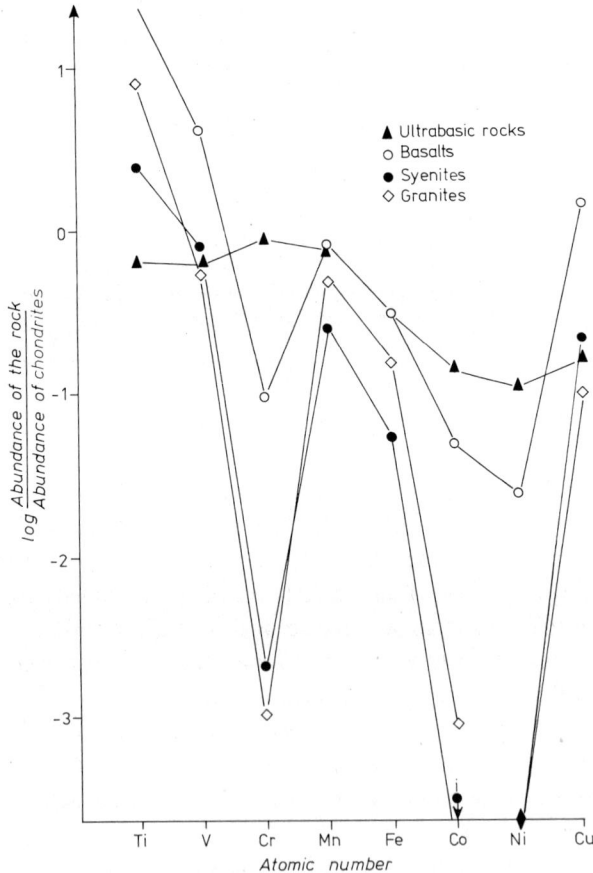

Fig. 4.10.   Abundances of the transition elements in the principal igneous rocks.

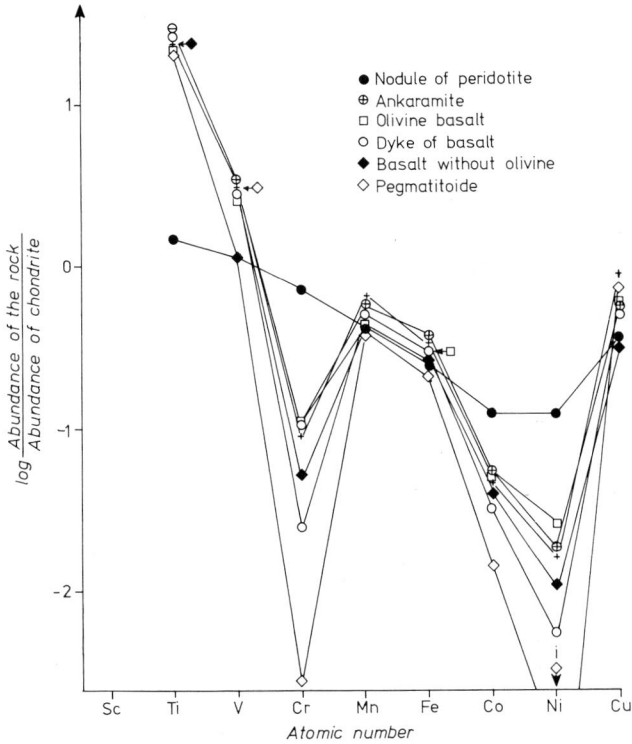

Fig. 4.11. Evolution of the transition elements in a magmatic series.

line field. It is known that the transition ions are characterized by partially filled
d-shell orbitals. Now these d-orbitals have special forms (Figure 4-12) that are
different according to whether they have their axis of symmetry along an axis or
diagonal to the axes.

When the ion is isolated, all the d-orbitals have the same energy. When on the
other hand the ions are in the field of negative charges, for example, a polyhedron of
negative ions, the repulsive forces that exist between the orbital electrons and the
negative ions of the polyhedron separate the different d-orbitals energetically. If the

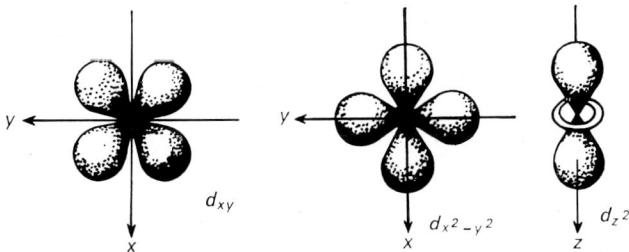

Fig. 4.12. Atomic d-orbitals.

negative-ion polyhedron is an octahedron, three orbitals are stabilized at the expense of the other two; if the polyhedron is a tetrahedron, two orbitals are stabilized at the expense of the other three.

Taking note of Δ, the energy difference between the two energy levels thus defined, and taking the mean energy as reference level, it can be shown that:

In the case of the enveloping octahedron, each one of the three stabilized orbitals has an excess energy of $-2\Delta/5$, and the two unstable ones have an excess of $+3\Delta/5$. In the case of the enveloping tetrahedron, each of the two stable orbitals carries a supplemental energy of $-3\Delta/5$ and each unstable orbital $+2\Delta/5$.

The transition ions have from 1 to 10 electrons, respectively, in d-orbitals. When they find themselves surrounded by the octahedral or tetrahedral structure, they acquire a stabilizing energy equal to the sum of the excess energies of the corresponding electrons.

One can then calculate for each one an energy of stabilization for the octahedral or tetrahedral surrounding and plot this energy as a function of atomic number (Figure 4-13). Now the transition ions enter into the essential minerals essentially in either octahedral positions (pyroxene, olivine) or tetrahedral ones (magnetite). It is thus theoretically possible to predict the various binding energies according to the sites of the various transition ions. The curves of stabilization energy vs atomic number in an octahedral field have the shape of an M (an inverted W). One can thus imagine the following simple model. A magma crystallizes. The solid contains a large number of octahedral sites. The transition elements enter into these octahedral sites. The partition coefficients (which, don't forget, are exponential functions of the binding energy) of the transition elements reflect these stabilization energies. As the magma

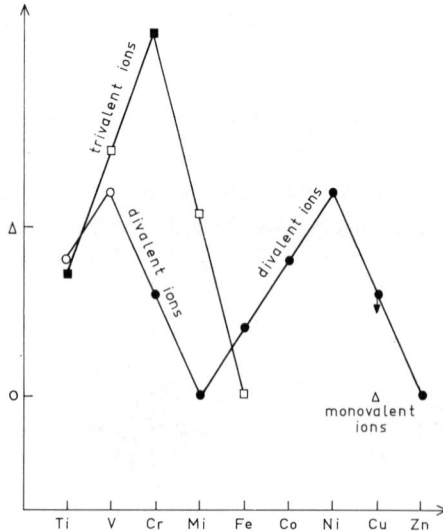

Fig. 4.13.   Energy of stabilization of ions of the transition elements in an octahedral envelope.

grows progressively poorer in the most stable elements in the solid (Cr and Ni for example), its distribution curve takes on a progressively sharper W form.

It is thus seen that the observed distributions show that in terrestrial magmatic differentiation, it is the distribution of octahedral sites of the solid magma that governs the geochemistry of the transition elements. To go farther and calculate quantitative models as for the rare earths, the partition coefficients must be known between minerals and between minerals and magma. Such work is in progress, but one can predict already that for basic magmas the results will usefully complement those obtained from the rare earths.

Here again, we have sought to follow a route which this time is based on an atomic and quantitative interpretation of the phenomena. We are approaching a veritable atomic geochemistry.

4.2. Trapping of Trace Elements in Sediments
(Metallogenic Importance of the Phenomena of Entrainment)

4.2.1. Impossibility of Direct Precipitation from Sea Water for Many Elements

It is well known that the number of abundant minerals in the present day ocean bottoms is quite small. In addition to the calcium carbonates, and the oxides of Fe and Mn, phosphates of Ca, and Ba sulfate are found locally but no trace of minerals of Zn, Cu, Pb, etc. It seems thus that for most elements, direct precipitation is impossible.

This state of affairs can be explained rather readily. In effect, sea water is a well defined medium and the concentrations of negative ions that can precipitate positive metallic ions is reasonably well known. The complexes formed by these metals with the elements of sea water are beginning to be known. If an uncertainty exists concerning activity coefficients, we shall see that it is certainly negligible. The problem reduces first to seeking the least soluble compound for each element and calculating the concentration of the element in the sea water in equilibrium with the solid. The ratio of the real, observed concentration to the calculated one constitutes what may be called the degree of saturation. Table 4-II summarizes the results established by Krauskopf (1956) and Sillen (1961).

Despite the lack of precision in the calculations, it can be seen at once that the elements V, Cr, Co, Ni, Cu, Zn, and Pb are very far from saturation, which explains the absence of specific minerals. Oppositely, the oxides of Fe and Mn are highly saturated and correspond to abundant minerals.

4.2.2. Precipitation in Anoxic Environments

In closed basins that are called 'anoxic' and stagnant, the abundance of organic debris creates a reducing medium where important amounts of H_2S can accumulate. Strøm has measured the total sulfide content ($\Sigma S = H_2S + HS^- + S^{2-}$) at 5×10^{-3} M in Norwegian fjords. The same order of magnitude is found in the interstitial waters of sediments. With the aid of these data, the maximum contents of Zn, Co, Ni, Cu,

TABLE 4-II

Sea water elements

Elements	Preponderant ionic form	Insoluble compound	Degree of saturation
Mg	Mg^{++}	$MgCO_3, H_2O$	3×10^{-2}
Ca	Ca^{++}	$CaCO_3$	$0.8-4$
Sr	Sr^{++}	$SrCO_3$	0.5
Ba	Ba^{++}	$BaSO_4$	$0.5-1$
V	$H_2VO_4^-$	V_2O_5, nH_2O	$<10^{-4}$
Cr	$CrO_4^=$	$CaCrO_4$	$<10^{-6}$
Mn	Mn^{++}	MnO_2	10^{+8}
Fe	$Fe(OH)_2^+, Fe(OH)_4^-$	Fe_2O_3	$10-1000$
Co	$CoCl^+$	$CoCO_3, Co(OH)_3$	$5 \times 10^{-7}-4 \times 10^{-6}$
Ni	Ni^{++}	$Ni(OH)_2$	$2 \times 10^{-7}-2 \times 10^{-5}$
Cu	$Cu^{++}, CuCl^+$	$Cu_2(CO_3)(OH)_2$	$10^{-3}-3 \times 10^{-2}$
Zn	Zn^{++}	$ZnCO_3$	$2 \times 10^{-3}-2 \times 10^{-2}$
Pb	$PbCl^+$	$PbCO_3$	$5 \times 10^{-3}-4 \times 10^{-2}$

Pb, Hg, and Ag in solution can be calculated. These (sulfide) solubilities are extremely small and they act to eliminate these elements little by little from the sea water.

Nevertheless, it is notable that sulfides other than those of Fe are virtually never found in anoxic basins. Still, Brooks *et al.* (1968) showed that in the interstitial water of sediments quite rich in sulfides ($\sim 10^{-3}$ M), the preceding elements are present in amounts very greatly in excess of what the solubility product would predict. The reason for this disagreement is as yet not known.

In summary, even in sulfurous environments, none of the trace elements envisaged could precipitate alone in sea water.

4.2.3. Processes of entrainment

The elements discussed above can thus not precipitate alone, yet it is certain that they are not concentrating in the sea water, a point to which we shall return in the next chapter. They must therefore be eliminated somehow from sea water, and their elimination constitutes processes which we shall refer to collectively as *entrainment*. Among them, we distinguish, sometimes a little arbitrarily, coprecipitation with formation of solid solutions, adsorption, and ion exchange.

Coprecipitation with calcium carbonates can play an important role particularly for Zn and Sr and possibly other metals, that with barium sulfate for Sr and chromate (?), that with phosphates for U and some of the rare earths. In the following chapter, it will be shown how the importance of such a process can be tested. Coprecipitation of a trace ion C^+ in a compound AB, in which C^+ can be substituted for B^+, is a special case of partition of an element between two phases. From the equations for the solubility products of salts AB and AC one may deduce:

$$\frac{(AC)}{(AB)} = \frac{(C^+)}{(B^+)} \times \frac{K_{SAB}}{K_{SAC}}$$

(AC) and (AB) designating the activities in the solid phase, (C^+) and (B^+) the activities in solution. A partition coefficient D is defined between the concentrations:

$$D = \frac{[AC]}{[AB]} \times \frac{[B^+]}{[C^+]}.$$

The activity coefficients of the ions, that have the same charge, are neighbors in the sense that

$$D \simeq \frac{K_{SAB}}{K_{SAC}} \times \frac{1}{\gamma},$$

where γ is the activity coefficient of AC in the solid solution. D may be determined experimentally or calculated from the diagrams of demixing of solid phases AB and AC.

Coprecipitation with the different kinds of calcium carbonate plays an important role in the geochemistry of Sr and Zn. Coprecipitation of the ions Fe^{2+} and Mn^{2+} with calcite leads in sufficiently reducing environments to calcites that are rich in Fe and Mn (up to several percent).

The barium sulfate that is frequently found in lacustrine or marine sediments can have entrained Sr, K, ($HKSO_4$ in solution in $BaSO_4$), and chromate. Calcium phosphates concentrate U and the rare earths.

Adsorption is a surface phenomenon, the laws governing which, however empirical or statistical, bring out the great influence of active surfaces. As a consequence, this phenomenon has its entire importance in geochemistry with poorly crystallized compounds like the metallic hydroxides. Among these may be cited the case of MnO_2 which is the essential substance of the nodules that pave the great ocean depths. These nodules contain up to 1 or 2% of Ni, Co, Ba, and Pb that can be reasonably included by adsorption. The adsorption of the ion Mn^{2+} is an indispensable step in the oxidation to MnO_2 and hence for the growth of the nodules.

This adsorption can be reversible, and for the case where all the superficial sites are occupied, they serve completely to yield ion exchange. It is in fact what happens specifically in the clays that are near to positive alkali ions, i.e., they are veritable exchange resins. Laboratory experiments have shown that these exchanges occur rapidly (equilibrium is reached in a matter of tens of minutes). The affinity increases from Na^+ to Cs^+ for most of the clay minerals, and this explains why the ratios of K/Na, Rb/K, and Cs/Rb are less in sea water than in the Earth's crust.

TABLE 4-III

Ratios of various elements

	Sea water	Crust
Na/K	28	1.1
K/Rb	3000	250
Rb/Cs	400	40

References

Albaréde, F. and Bottinga, Y.: 1972, *Geochim. Cosmochim. Acta* **36**, 141.

Allègre, C. J., Javoy, M., and Michard, G.: 1968, in T. Ahrens (ed.), *Origin and Distribution of the Elements*, Pergamon Press, New York.

Burns, R. and Fyfe, S.: 1967, in P. Abelson (ed.), *Research in Geochemistry*, Vol. II, John Wiley and Sons, New York.

Coryell, C. D., Chase, J. W., and Winchester, J. W.: 1963, *J. Geophys. Res.* **68**, 559.

Gast, P. W.: 1968, *Geochim. Cosmochim. Acta* **32**, 1057.

Masuda, A.: 1962, *J. Earth Science* **10**, 173.

McIntyre, W. L.: 1963, *Geochim. Cosmochim. Acta* **27**, 1209.

Montigny, R., Bougault, H., and Allègre, C. J.: 1974, 'Trace Elements Contents and Genesis of Ophiolithe Complexes', to be published.

Neuman, H., Mead, J., and Vitaliano, C. J.: 1954, *Geochim. Cosmochim. Acta* **6**, 90.

Schilling, J. G. and Winchester, J. W.: 1974, 'Rare Earth Fractionation and Magmatic Processes in Mantle of Earth and Terrestrial Planets', to be published.

Schnetzler, C. C. and Philpotts, J. A.: 1970, *Geochim. Cosmochim. Acta* **34**, 331.

Shaw, D.: 1968, *Geochim. Cosmochim. Acta* **32**, 573.

Shaw, D.: 1970, *Geochim. Cosmochim. Acta* **34**, 237.

CHAPTER 5

IRREVERSIBLE PROCESSES OF ELEMENT TRANSFER

Introduction

In all that has gone before, it has tacitly been assumed that chemical equilibrium was attained. While to be sure certain processes at high temperatures do reach equilibrium in short time intervals, the great mass of natural reactions stop far short of states of equilibrium. This is especially the case for interactions between the lithosphere and the hydrosphere, and even more so those where the speed of flow of water over rocks is large.

Numerous highly different attempts to analyze natural processes have been envisaged:

– A thermodynamic way through study of irreversible processes based on the principle of partial equilibrium;

– a kinetic method of studying the combined diffusion and reaction chemistry; and

– a third way, that in some ways approaches the preceding one, that is an elegant method of calculating balances (method of boxes).

These three modes of thinking can clearly be applied independently to all geochemical problems. In any case, for certain problems, one or another approach is likely to prove more successful, if at first provisionally.

Without unduly extending the number of examples, we show here how weathering and metamorphism can be treated by irreversible thermodynamics, how the redistribution of certain chemical elements in the oceans can be explored by 'the box model', and finally we present an example of diagenetic evolution treated by the kinetic method.

In conclusion, and as an application of this chapter, we shall show how the transfer processes to the non-equilibrium states can lead to good quality chemical separations applicable on a global scale, as for example, the separation of manganese in nodule form in great ocean depths.

5.1. Evaluation of the Transfer of Elements During Irreversible Geochemical Processes

5.1.1. GENERALITIES

The study of problems of equilibrium of minerals has advanced greatly in recent years. Various researchers (especially Helgeson, 1968, 1969) have tried in addition to unify the 'principles' of equilibria of minerals and of the geochemistry of solutions and have tried to apply the results to phenomena of geological interest.

Among these, the interactions between silicates and alumino-silicates of rocks of deep origin and of aqueous solutions figure in the forefront. These processes include meteorological weathering, phenomena of metasomatism, etc. More precisely, these tentative attempts grow to prime importance each time that a complex mineral shows an incongruent solubility, that is, only one portion of the constituent elements is soluble.

Since a mineral is in general unstable in the presence of natural solutions, the global process of transference of elements from the solid phase to solution is irreversible. But the 'principle' of local equilibrium leads to the picture that global processes evolve from a succession of partial equilibrium states. Thus the reaction of a mineral with a solution in which it is not in equilibrium leads to the formation of various complex ions in the solution that are in equilibrium among themselves and to the creation of new minerals that are in equilibrium with the solution. The changes in composition of the different phases depend on the path chosen. The objective of what follows is to foresee the 'path' of the irreversible reaction.

5.1.2. MOBILE AND INERT ELEMENTS. (GRAPHICAL METHOD OF DISCUSSION)

In considering primarily the case of the silicates and alumino-silicates, the component elements of the minerals can be divided into three classes:

(1) The soluble positive ions that can be liberated in solution where their solubility is unlimited. Examples are Na^+, K^+, Ca^{2+}, Mg^{2+}, and sometimes Fe^{2+}. It should be noted that these examples are only indicative and that depending on the case, one or more of these ions will not meet the conditions of the definition.

This first class of elements will be called the mobile elements.

(2) The mobile elements that are simultaneously present in the mineral and in one phase in addition. This is always the case for H^+ and most often for Si.

(3) Finally, the elements that are present in solution in extremely small amounts, that are characterized as the inert elements, for example, Al.

Korjinski has shown that the study of complex assemblages of minerals could be elegantly carried out starting from diagrams that introduce the chemical potentials of the mobile elements. Figure 5-1 presents an example of equilibrium between K and Mg Al silicates, the diagram showing $\log([Mg^{2+}]/[H^+]^2)$ as ordinate against $\log([K^+]/[H^+])$ as abscissa, in the presence of quartz, at $300\,°C$.

It is further possible to show that:

(1) When the solution is in equilibrium with a single mineral, the point that traces the evolution of the system in such a diagram describes a straight line whose slope is equal to the ratio of the charges of the mobile elements considered, provided that the concentration of H^+ is low enough.

(2) If starting from the solution, more than one mineral deposits, the representing point in the graph describes the curve that limits the domain of stability of these minerals, and this imposes a supplementary condition on the determination of the

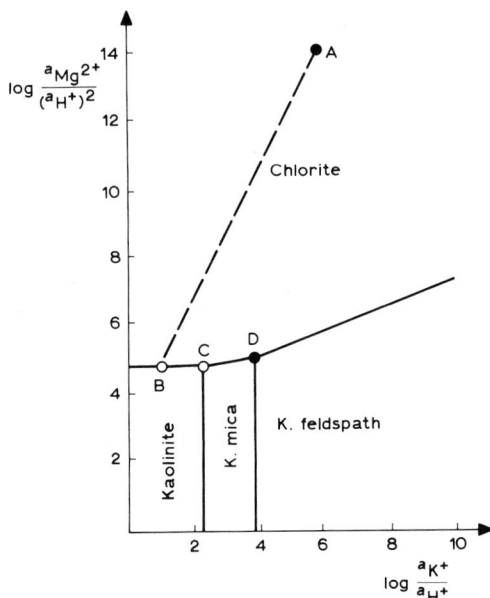

Fig. 5.1. Korjinski diagram.

stoichiometric coefficients of the reaction. We shall return to this point in more detail in the following paragraph.

By way of example, let us consider the following problem: A potash feldspar at 300 °C and a suitable pressure whose influence we ignore here is in contact with an aqueous solution whose composition is near to that of sea water. In order not to complicate the problem too severely, we confine ourselves to the most important exchange that concerns Mg, for the sea water corresponds to a domain of stability of chlorite $Mg_5Si_3Al_2O_{10}(OH)_8$, and the K in the feldspar according to the following reaction:

$$2KAlSi_3O_8 + 5Mg^{2+} + 8H_2O$$
$$\rightarrow Mg_5Al_2Si_3O_{10}(OH)_8 + 8H^+ + 2K^+ + 3SiO_2.$$

The solution being in equilibrium with the only mineral, chlorite, its trace point in the diagram (Figure 5-1) moves along a line of slope 2 up to point B where the domain of stability of kaolinite is reached. From there on, the action of the sea water on the feldspar leads to the simultaneous formation of chlorite and kaolinite. The coefficients of the reaction

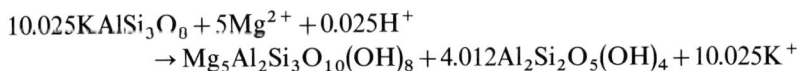

$$10.025KAlSi_3O_8 + 5Mg^{2+} + 0.025H^+$$
$$\rightarrow Mg_5Al_2Si_3O_{10}(OH)_8 + 4.012Al_2Si_2O_5(OH)_4 + 10.025K^+$$

are determined by the conservation of the elements and by the supplementary condition that the composition of the solution must vary along BC.

At C, there is destruction of the kaolinite to the profit of potassium mica:

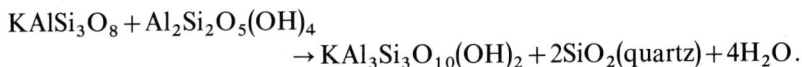

$$KAlSi_3O_8 + Al_2Si_2O_5(OH)_4$$
$$\rightarrow KAl_3Si_3O_{10}(OH)_2 + 2SiO_2(quartz) + 4H_2O.$$

Thereafter, the solution allows chlorite and mica to deposit and evolves along CD. The reaction is written:

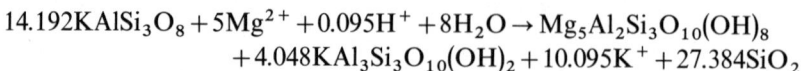

$$14.192KAlSi_3O_8 + 5Mg^{2+} + 0.095H^+ + 8H_2O \rightarrow Mg_5Al_2Si_3O_{10}(OH)_8$$
$$+ 4.048KAl_3Si_3O_{10}(OH)_2 + 10.095K^+ + 27.384SiO_2$$

and the solution evolves to D where it is in equilibrium with feldspar and the reaction stops. The graphical method of Helgeson has permitted us to determine by what route the equilibrium between feldspar and sea water is reached. It also permits us to see that the evolution occurs with the destruction of 0.14 mole of feldspar and the formation of 0.01 mole of chlorite, 0.28 mole of quartz, and 0.04 mole of mica per kilogram of water.

5.1.3. SECOND METHOD OF HELGESON

The graphical method has the obvious advantage of simplicity. Against it is that it does not lead to identification of all the problems. Thus the movement of the point in the Korjinski diagram can be foreseen only in special cases. Moreover, it does not take into consideration an important group of compounds, namely, the soluble complexes. Helgeson (1968) accordingly returned to the problem with the same hypotheses for which he proposed a matrix solution.

If, for example, the reaction of kaolinite with water is studied, and all ions, complex ions, and insoluble products that may form are accounted for, a reaction equation is reached having the following form:

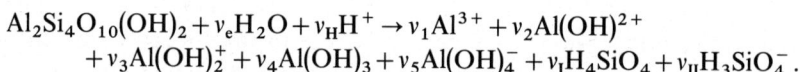

$$Al_2Si_4O_{10}(OH)_2 + v_eH_2O + v_HH^+ \rightarrow v_1Al^{3+} + v_2Al(OH)^{2+}$$
$$+ v_3Al(OH)_2^+ + v_4Al(OH)_3 + v_5Al(OH)_4^- + v_lH_4SiO_4 + v_{ll}H_3SiO_4^-.$$

The parameters v change throughout the course of the reaction in such a way as to satisfy the following two types of conditions:
(a) The elements are conserved;
(b) equilibrium exists among all the solids and the complexes that form and the ions of which they are composed.

This equilibrium is displaced quasi-statically in accordance with the law of mass action, the solution evolving so as to remain in balance as the reaction advances by an amount $d\xi$.

$Al(OH)^{2+}$, Al^{3+}, and H^+ must balance at every instant:

$$\frac{(Al(OH)^{2+})(H^+)}{(Al^{3+})} = K.$$

The changes in these quantities must specifically satisfy

$$\frac{\frac{d}{d\xi}(Al(OH)^{2+})}{(Al(OH)^{2+})} + \frac{\frac{d}{d\xi}(H^+)}{(H^+)} - \frac{\frac{d}{d\xi}(Al^{3+})}{(Al^{3+})} = 0$$

and if one supposes that the activity coefficients of each ion remain constant during an infinitesimal advance of the reaction:

$$\frac{\frac{d}{d\xi}[Al(OH)^{2+}]}{[Al(OH)^{2+}]}+\frac{\frac{d}{d\xi}[H^+]}{[H^+]}-\frac{\frac{d}{d\xi}[Al^{3+}]}{[Al^{3+}]}=0.$$

Now by definition of the degree of reaction, each numerator is equal to the stoichiometric coefficient of the corresponding compound in the reaction. Thus the equation

$$\frac{v_2}{[Al(OH)^{2+}]}+\frac{v_H}{[H^+]}-\frac{v_1}{[Al^{3+}]}=0$$

is linear in the v's for each equilibrium and combines with the equations of conservation of elements which are also linear in the v's. It is easy to see that the number of equations is equal to the number of unknowns. In effect, for each element for which a conservation equation is written, one and only one ion can be made to correspond, a simple one if it exists, otherwise an arbitrarily chosen complex ion. One can also write a law of mass action (or an expression derived therefrom) for each complex and each insoluble compound.

Thus if at one instant in the evolution of the reaction, the molalities of each compound in solution are known, the coefficients v of the reaction for the next infinitesimal step characterized by $\Delta\xi$ can be determined by solving a system of linear equations. Evaluation of all the v's permits determination of the molalities of each compound at the end of this step:

$$m_i=m_i^\circ+v_i\Delta\xi.$$

It is thus possible to follow the progress of the reaction step by step.

By way of example, the case may be considered of the simultaneous weathering of potash and soda feldspars with water (Helgeson, 1969) under the assumption that the weathering rates are equal for the two feldspars. The simplified form of the reaction equation at its initiation is

$$KAlSi_3O_8+NaAlSi_3O_8+16H_2O\rightarrow 2Al(OH)_4^-+K^++6H_4SiO_4$$

and the reaction path may be plotted on diagrams with $\log a(H_4SiO_4)$ as abscissa and either $\log(a_{Na^+}/a_{H^+})$ or $\log(a_{K^+}/a_{H^+})$ as ordinate (Figure 5-2). In addition, the masses of solid compounds formed or destroyed during the reaction are shown in Figure 5-2b.

The first solid phase to deposit is gibbsite (A in Figure 5-2b) followed at B by the first kaolinite formation at the expense of the gibbsite. At C, the gibbsite has completely vanished. Thereafter, at D, the potassium mica deposits, and then at E the solution reaches equilibrium with potash feldspar. But albite continues to react with the solution liberating Si and Al, as a result of which mica is first transformed to potash feldspar and that then precipitates. At F, the mica has totally vanished and the depositing minerals are kaolinite and feldspar. Continuing, at G, the limit kaolinite-

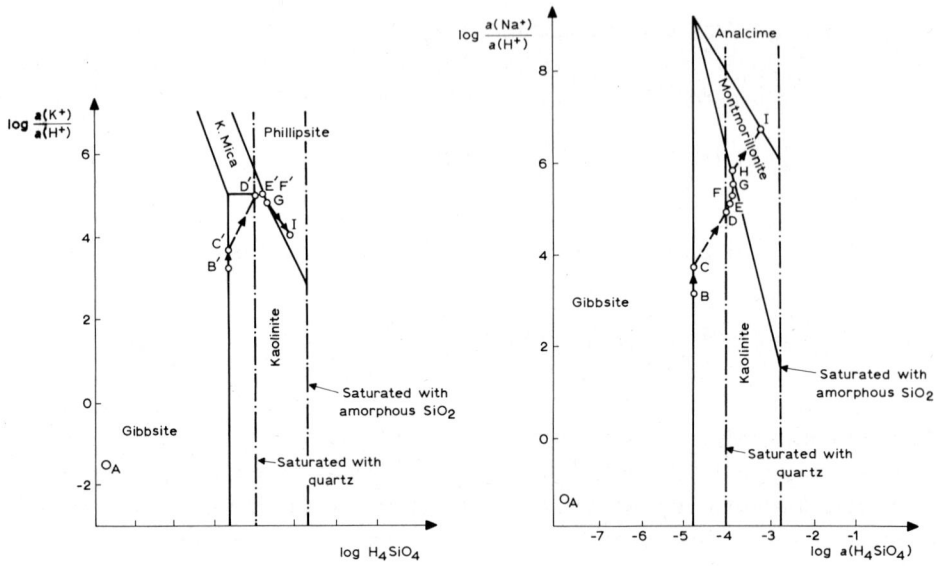

Fig. 5.2a.   Korjinski diagram of alkaline feldspars.

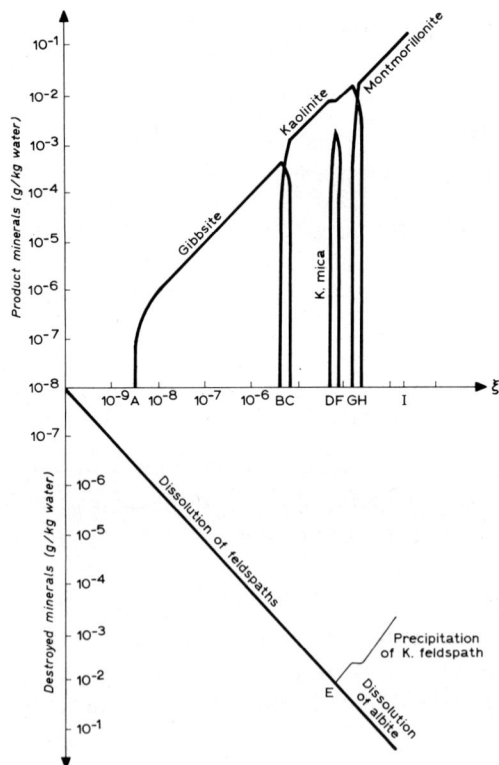

Fig. 5.2b.   Mineral formation during weathering of alkaline feldspars.

montmorillonite is reached, and the latter mineral forms at the expense of kaolinite which is consumed at H. The last phase of the process leads to deposit of feldspar and montmorillonite until final equilibrium is reached at I when the solution is in equilibrium with potash feldspar, albite, and montmorillonite.

In an open system, the aqueous phase leaves the reacting mineral before global equilibrium is reached. The transfer of elements is therefore different, but the combinations of authigenic minerals will be the same whether the system is open or closed. The association that we have encountered in this study is typical of weathering horizons.

The methods of Helgeson are still too new to have been subjected to applications in the field by geologists. But though many points of detail remain to be studied, such as the thermodynamic constants, compositions of minerals, extension to open environments..., the method introduces for the first time the chemical principles making a quantitative study of the phenomena of metasomatism and weathering possible.

5.2. Box Model

5.2.1. CONCEPT OF RESIDENCE TIME

The box model developed chiefly by Craig (1969) and Broecker *et al.* (1960, 1961, 1962) consists of dividing the complete object of study, here the ocean-atmosphere system, into a certain number of parts, in which the various quantities under study (T, concentrations,...) are constants in space and time. The system is thus pictured as in a stationary state. The method of boxes has as its goal the estimation or calculation of the transferences of materials among boxes.

A most interesting concept that emerges from this model is that of residence time. Consider for example a 'box' containing a quantity Q of a certain element A

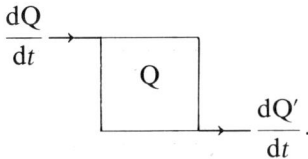

$$
\frac{dQ}{dt} \rightarrow \boxed{\quad Q \quad} \quad \frac{dQ'}{dt}.
$$

In unit time, a quantity dQ/dt of this element enters the box and a quantity dQ'/dt leaves. The assumption that the state is stationary implies that

$$
\frac{dQ}{dt} = \frac{dQ'}{dt}.
$$

The mean time that the element A resides in the box is

$$
\tau = \frac{Q}{dQ/dt}.
$$

It is this quantity τ that is called the residence time of element A in the box under study. This concept was first applied to the ocean as a whole.

It has been possible to estimate the amount of each of the elements reaching the sea per unit of time and also the amounts of these elements leaving the environment by precipitation during the same time. These quantities proved to be sensibly equal, justifying the assumption of stationarity. It was thus possible to calculate the 'global' residence times of all the elements in sea water (Barth, 1956; Goldberg and Arrhenius, 1958).

Progress in the estimation of the contents of streams and rivers has permitted evaluation of the quantities of elements reaching the sea in a dissolved state and to calculate a *residence time for the soluble elements* in the ocean, the meaning of which is clearer than the preceding calculation (Wedepohl, 1968). In every case it is necessary to make a so-called 'salt-cycle' correction for those elements that are transported from the ocean to the land by rain water. This correction is especially important for the halide ions.

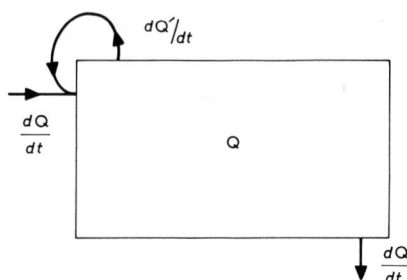

Table 5-I shows the calculated values of residence times for certain elements.

Although the values are often only indicative because the estimates of mean values are often marred by gross uncertainties, it can be seen that the agreement is satisfactory for the most soluble elements that are obviously those that have the longest residence times.

The residence time of a trace element in sea water may permit testing of the importance of a mechanism of precipitation (Michard, 1971). If one may picture that an element C is eliminated from sea water by entrainment with an element B in a compound AB, one must in effect have:

$$\frac{dC/dt}{dB/dt} = D_{CB} \frac{C}{B}$$

(D_{CB} being the partition coefficient between B and C, between the mineral AB and water) and this leads to:

$$\tau_B = D_{CB}\tau_C.$$

If a more effective phenomenon intervenes, one has:

$$\tau_B \gg D_{CB}\tau_C.$$

TABLE 5-I

Residence time

Element	Residence time of the soluble element	Global residence time
Na	2.6×10^8	2.6×10^8
Mg	4.8×10^7	4.5×10^7
K	2.4×10^7	1.1×10^7
Sr	1.9×10^7	1.9×10^7
Ca	3.2×10^6	8×10^7
U	4.3×10^5	5×10^5
Ni	1.6×10^5	1.8×10^4
Zn	1.4×10^5	1.8×10^5
Mn	4.5×10^4	1.4×10^3
Cu	4.5×10^4	4.5×10^4
Ti	2.6×10^4	160
Si	2.4×10^4	8×10^3
Al	4.5×10^3	100
Fe	4×10^4	140

Unfortunately, these estimates are at the moment too inaccurate for drawing of definitive conclusions. For example, for Mn in calcite, one has

$$\frac{\tau_{Ca}}{\tau_{Mn}} = 70, \qquad D_{CaMn} = 5$$

and it thus seems that there is a trapping mechanism for the Mn from the sea water that is more effective than the coprecipitation with calcite. In the case of Zn $((\tau_{Ca}/\tau_{Zn}) = 20, D_{CaZn} = 5)$ it is too early to draw the same conclusion.

5.2.2. MODELS OF THE OCEAN

The box models presented by Broecker *et al.* (1961) to describe the oceans of the Earth employ a radioactive tracer to measure the residence time. We shall demonstrate the principle with the simplest case: The ocean is divided into two parts, an upper zone that is warm and is well mixed, and a deep zone that is cold and stratified. The two parts are readily distinguished by their concentration of carbon and particularly sharply by that of ^{14}C (Figure 5-3).

The exchanges of the element C between the two marine reservoirs take place largely in dissolved forms but partially by a flux of particles.

We then write:

– B is the flux of solid particles in mole yr^{-1}.

– W and W′ are the volumes of sea water descending from the upper zone to the lower, in $m^3 yr^{-1}$ and ascending from the lower to the upper.

– C_s and C_d are the concentrations of C dissolved in the shallow and the deep zones, respectively.

– R_s and R_d are the ratios of $^{14}C/C$ in the respective zones.

– V_d is the volume of the deep zone.

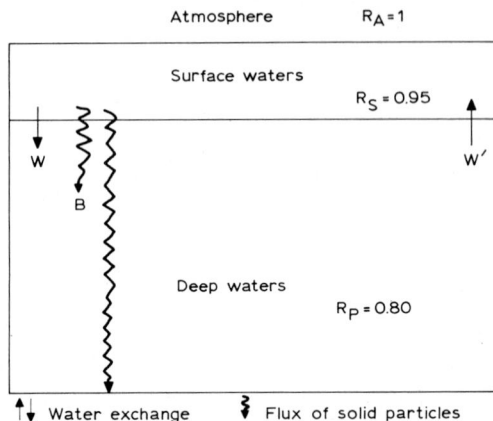

Fig. 5.3.   The box model.

By the hypothesis of steady states, the conservation of water necessitates that

$$W = W',$$ (1)

– for total C content

$$(WC_s + B) = WC_d$$ (2)

– for ^{14}C

$$(WC_s + B) R_s = WC_d R_d + V_d C_d R_d \lambda$$ (3)

where λ is the radioactive decay constant of ^{14}C.
The residence time in the deep zone is

$$\tau_d = \frac{V_d C_d}{WC_d} = \frac{V_d}{W}$$

which, on combining with Equations (2) and (3) gives

$$\tau_d = \left(\frac{R_s}{R_d} - 1\right)\frac{1}{\lambda}.$$

Now $\lambda = 1.25 \times 10^{-4}$ yr^{-1} so $\tau_d = 1200$ yr.
Using a large number of analyses of ^{14}C in the oceans, Broecker *et al.* (1960, 1962) were led to distinguish within the global ocean six large reservoirs whose characteristics are displayed in Table 5-II.
The simplifying assumptions on the systems of currents that cause the mixing of water are shown in Figure 5-4.
The transference of water across the thermoclines in both the Pacific and Atlantic

TABLE 5-II

Characteristics of reservoirs

No.	Geographic site	Latitude	Depth (m)	Area (m^2)	Quantity of CO_2 (mol)	$^{14}C/^{12}C$ atm $^{14}C/^{12}C$
2	Pacific and Indian Ocean surface water	$<50\,°S$	<100	220×10^{12}	50×10^{15}	0.97
3	Atlantic Ocean surface water	$55\,°N–50\,°S$	<100	80×10^{12}	18×10^{15}	0.97
4	Pacific and Indian Ocean deep water	$<50\,°S$	>100		1980×10^{15}	0.80
5	Antarctic	$>50\,°S$		45×10^{12}	420×10^{15}	
6	Atlantic deep	$55\,°N–50\,°S$	>100		655×10^{15}	0.90
7	Arctic	$>55\,°N$		15×10^{12}	40×10^{15}	

Oceans is assumed to be negligible, as well as the transfer between the Arctic and the Pacific through the Bering Straits.

The interchange of CO_2 between the atmosphere and the ocean is supposed to be the same at all points of the Earth.

The Atlantic Ocean is characterized by northward surface currents that are compensated by oppositely directed motion at greater depths.

If one writes that the ^{14}C is stationary in each of these reservoirs, the flux of CO_2 between the atmosphere and the ocean and the flux of water between each of the reservoirs can be calculated. It is also possible to obtain the residence times of carbonic gas and of the substances dissolved in each reservoir (Table 5-III).

The authors have in addition recomputed these values by modifying some of their hypotheses such as the thickness of the surface zone, varying speeds of transfer of

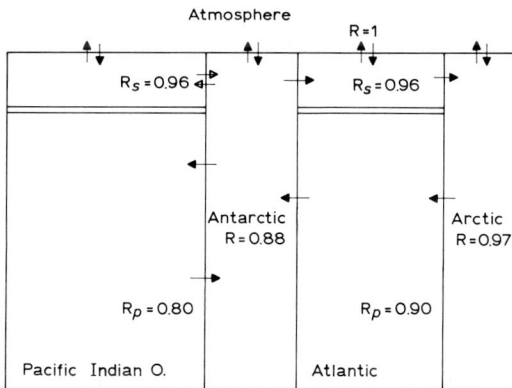

Fig. 5.4.   Model for the world's oceans (after Broecker et al., 1961).

TABLE 5-III

Residence times

	Residence times (yr)	
	CO_2	Dissolved substances
Atmosphere	7	
Pacific + Indian (surface)	8	30
Atlantic (surface)	10	20
Pacific + Indian (bottom)	800	800
Anarctic	70	70
Atlantic (bottom)	650	650
Arctic	30	40

CO_2 between the atmosphere and the various reservoirs, etc. The results are only very slightly altered.

It is of course clear that this problem of mixing times of ocean waters has a considerable interest outside the field of oceanographic chemistry. (The comparison of the residence times of the elements with those of the waters discloses that certain elements like Fe and Al cannot be regarded as homogeneously distributed throughout the ocean.) Interest in mixing times extends to marine biology and also to the dilution of radioactive waste products and other industrial 'poisons'. Thus the liberation of Pb into the atmosphere by motor vehicles that has grown during the past few decades leads to an accumulation of this metal in surface layers, as appears in the profile of Pb distribution in the Pacific Ocean (Patterson-Tatsumoto, Figure 5-5).

If the quantity of Pb released annually into the atmosphere remained stationary, it would take six centuries before the distribution of Pb in the ocean became homogeneous.

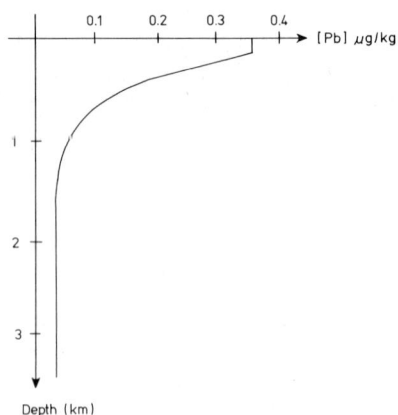

Fig. 5.5.   Concentration of dissolved Pb in sea water as a function of depth (after Patterson and Tatsumoto).

5.3. Chemical Diffusion-Reaction Coupling

5.3.1. GENERALITIES

The dissolved elements in natural solutions undergo motions that may be separated into two categories.

In one category are the movements caused by the aggregate motion of the solution that can preponderate, as in rivers and streams for example. The concentration of an element is then a function of space and time. We name such movement *advection*. The conservation of an element in a volume surrounding a point under consideration is

$$\frac{\partial C}{dt} = -v\frac{\partial C}{\partial x},$$

where v is the speed (for a unidimensional problem).

In the other category are the spontaneous motions of the ions in a solution tending to reestablish the equality of chemical potentials at all points in a volume. This is the phenomenon of *diffusion*. In the case of a unidimensional diffusion, the fundamental equation, called Fick's law, may be written

$$\frac{\partial C}{\partial t} = D\frac{\partial^2 C}{\partial x^2},$$

where D is the coefficient of diffusion and is dependent on the nature of the ion and the solution. The other causes of variations in concentrations with time are chemical reactions, and for some elements radioactive disintegration. Chemical reactions may be of various orders; however, in the most important cases, the consumption of O_2 and production of CO_2, the amount consumed or produced depends only on one variable – the quantity of organic material oxidized – which is independent of the concentrations of the key elements or substances, O_2 and CO_2, in solution. In most other cases, first order kinetics gives a good approximation. Finally, radioactive disintegration obeys a law similar to that for a first order chemical reaction, so that a general equation covering all mechanisms takes the form:

$$\frac{\partial C}{\partial t} = D\frac{\partial^2 C}{\partial x^2} - v\frac{\partial C}{\partial x} - \lambda C + J,$$

where λ combines all first-order terms, chemical reactions, and disintegrations, and J includes all terms of order zero.

In most cases, only steady state phenomena are envisaged, so that the equation reduces to

$$D\frac{\partial^2 C}{\partial x^2} - v\frac{\partial C}{\partial x} - \lambda C + J = 0.$$

Numerous examples of the application of this equation can be cited, either to the grand scale, slow motions of the ocean (Craig, 1969 – the author shows that the equation applies if D is replaced by another coefficient called the eddy diffusion coefficient), or to the diagenetic evolution of a sediment. Among the numbers of cases treated, we select here a model of the distribution of sulfates in the interstitial water of sediments recently used by Berner (1964) to study the deposits in the Santa Barbara basin of the California coast.

5.3.2. DISTRIBUTION OF SULFATES IN THE INTERSTITIAL WATERS OF THE SANTA BARBARA BASIN

The sulfates are reduced to sulfides in the interior of the sediment by a bacterial reduction, the bacterium *Desulfuvibrio desulfuricans* being the principal active agent. Berner believes that the speed of the reduction is proportional to the concentration G of the organic material that can be utilized by the bacteria.

The reduction reaction can be schematically written:

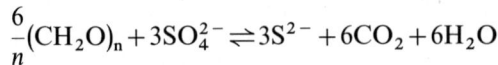

$$\frac{6}{n}(CH_2O)_n + 3SO_4^{2-} \rightleftharpoons 3S^{2-} + 6CO_2 + 6H_2O$$

and its rate is:

$$\frac{\partial C}{\partial t} = -\alpha KG$$

K being the rate constant and α the ratio of the stoichiometric coefficients of the sulfate and CO_2 in the oxidation-reduction reaction (thus in the preceding schematic example, $\alpha = \frac{1}{2}$).

For the organic material in suspension, diffusion does not enter, and the general equation reduces to

$$\frac{\partial G}{\partial t} = -v\frac{\partial G}{\partial x} - KG.$$

If the stationary state is reached, this reduces further to:

$$\frac{\partial G}{\partial t} = 0 = -v\frac{\partial G}{\partial x} - KG$$

and

$$G = G_0 \exp\left(-\frac{K}{v}x\right).$$

For the sulfates, the general equation then is

$$D\frac{d^2C}{dx^2} - v\frac{dC}{dx} - KG_0 \exp\left(-\frac{K}{v}x\right) = 0$$

which integrates readily to give

$$C = C_0 - \frac{G_0}{\frac{DK}{v^2} + 1}\left(1 - \exp\left(-\frac{K}{v}x\right)\right).$$

The results of Kaplan (1963) on the Santa Barbara Basin apply simultaneously to concentration of the sulfates, shown in Figure 5-6 to organic material whose abundance in the dry soil is 3% in the surface zone ($0 < x < 50$ cm) and 2% below 3 m, which permits evaluation of the amount of organic matter available, and the velocity of sedimentation which is of the order of 0.3 cm yr^{-1}.

The distribution of the sulfates is adequately represented by the equation

$$C = 27 - 19[1 - \exp(-0.015x)]$$

and this enabled Berner to calculate K as 1.4×10^{-10} s^{-1} and D as 0.3×10^{-5} cm^2 s^{-1}.

The value of K is of the same order of magnitude as that calculated by Kaplan from the localization of virtually all of the pyrite in the first ten centimeters of sediment.

The value of D is quite comparable with that obtained experimentally for chloride ions in water and sediment mixtures (Priklonskii et al., 1959).

Thus although this sort of model represents a great simplification of nature, it does seem to permit a logical approach to it.

Otherwise, the data on the sulfides in the Black Sea can also be treated in the same way, and the linear relations can be derived connecting sulfate, carbonate, and sulfide, seen experimentally by Skopentsev et al. (1958).

This type of study is of definite interest for all elements subject to a change of degree of oxidation in the sediment. Some quantitative models have been elaborated for U (Ku, 1966) in connection with the measurement of age by radioactive decay. Others have been made for Mn that concentrates in the upper oxidized zone, by diffusion from deep reducing sediment, where Mn^{2+} is remobilized (Lynn and Bonatti, 1965; Michard, 1971). There is, in this area, an interesting approach to the phenomena of strong diagenesis concerning which the chemistry is poorly known.

5.3.3. DIFFUSION AND PARTITION COEFFICIENT

During the crystallization of a mineral either from a solution or a melt, the mineral can incorporate trace elements by substitution in place of a primary element of the

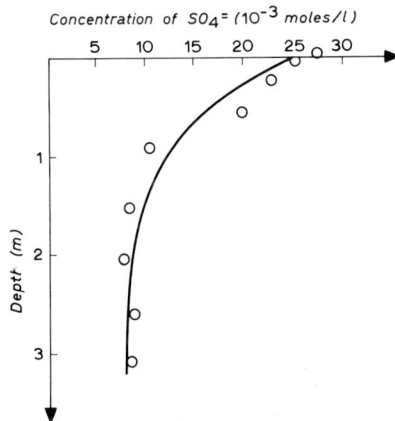

Fig. 5.6.   Concentration of dissolved sulfates in the sediments of the Santa Barbara Basin (after Kaplan, 1963).

crystal. The partition at equilibrium takes on a certain value K. If K is, for example, very large compared wich unity for some trace element designated Tr, the solution or melt in contact with the crystal becomes severely depleted in Tr, and if the rate of diffusion of Tr is inadequate compared with the rate of crystallization, the incorporation of Tr is less than under equilibrium conditions. Albarède and Bottinga (1971) have studied the process numerically.

The results can be interpreted by plotting the ratio of concentration in the crystal to concentration in the liquid as a function of the distance from the interface (Figure 5-7).

Three cases may be represented:

(a) The crystal is in universal equilibrium with the solution (Berthelot-Nernst law);

(b) The surface of the crystal is in equilibrium with the solution (Doerner-Hoskins law);

(c) The replacement of the trace element in the liquid near to the crystal is limited by diffusion (model of Albarède and Bottinga).

It thus is clear that one can measure the partition coefficient and *true* D only in two cases, (a) and (b) above:

– Using the mean concentration in the solid where the partition is of the Berthelot-Nernst type;

– Using the surface concentration in the crystal (something that may be experimentally difficult to do) if there is a Doerner-Hoskins distribution.

If D1, the apparent coefficient measured on analyzing the crystal ensemble is greater than D (curve b′) the distribution is of Doerner-Hoskins type, and if less than D (curve c′) the Albarède-Bottinga model is applicable (Figure 5-3).

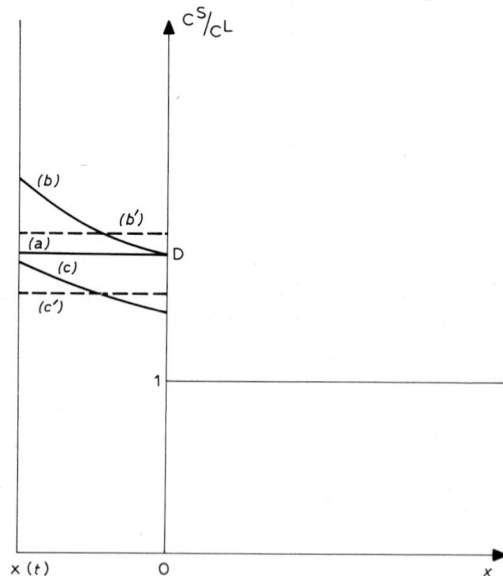

Fig. 5.7.   Concentration of a trace element in a crystal growing from a melts.

5.4. Deposit of Manganese in Ocean Depths

We wish with this example to try to show the practical importance that can be assumed by the transference of elements in a non-equilibrium state.

Everyone by now has heard of the nodules of Mn that cover an important fraction of the great ocean depths and in which there is considerable practical interest, not only as a source of Mn but especially as a source of Co and Ni. These nodules are formed by concentric layers of a mixture of MnO_2 and oxides of Fe (trivalent) surrounding a nucleus of variable nature, limestone, silica, etc.

One of the chief problems posed by the existence of the nodules is to explain how such a chemical separation could occur in natural waters.

If one examines the degrees of saturation of the elements in sea water (cf. Chapter 4) one can see at once why it is Mn and not another element that can cause this sort of separation. In effect, it attains a degree of saturation of 10^8 whereas that of Fe, which is one of the rare other elements to be supersaturated, is only of the order of 10. The reaction:

$$\tfrac{1}{2}O_2 + Mn^{2+} + H_2O \rightarrow MnO_2 + 2H^+$$

would lead, at equilibrium, to an activity of dissolved Mn equal to 10^{-16} M l^{-1}, and the measured value is of the order of 10^{-8} M l^{-1}. But at these last concentrations, the rate of the reaction is infinitesimally slow. All solid surfaces serve as catalyzers, and in particular among the natural substances, $CaCO_3$ and hydroxides of Fe and of Mn. Examination of the solubility diagram of Fe and Mn as a function of pE at pH 8 shows that the Fe follows a different behavior pattern. For values of pE already much less than the measured potentials, Fe is trivalent, and thereafter, its solubility, which is small but not zero, no longer changes as pE increases. The solubility of the Mn by contrast never stops decreasing. One can thus imagine the separation of Fe and Mn occurring in the following way:

Iron, which is relatively rapidly oxidized by O_2, is immobilized relatively near to its weathering site. That portion that reaches the sea is oxidized relatively rapidly there and deposits in coastal zones. The Mn by contrast oxidizes slowly and primarily at the water-sediment interface. Its concentration in sea water may be regarded as a stationary state between influx and deposition.

A new problem arises if the quantity of Mn deposited on the ocean floor per year is measured. It is distinctly greater than the amount delivered by streams and rivers. And this feature is once more peculiar to Mn. The solution to the problem is not to be sought by invoking special transports (submarine volcanism, for instance) but by using a special reaction peculiar to Mn. The study of cores of sediments extracted from ocean bottoms has shown that the concentration of Mn is generally greater at the top of the core. This is tied to a change in the conditions of oxidation; the top of the core is oxidizing, the bottom is reducing. Conversely, the concentration of Mn in interstitial water is very high in the deep-lying part of the core (as much as 1000 times that in sea water) and is virtually zero in the upper portion. One can thus believe that

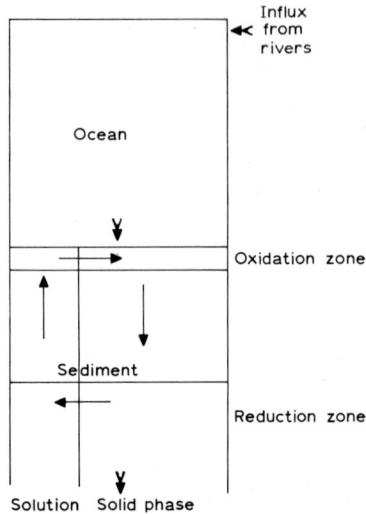

Fig. 5.8.   The Mn cycle in the ocean.

the 'excess' of Mn is furnished by Mn that moves upward, by diffusion in the dissolved phase, just where its concentration is the weakest. The following scheme is thus reached, shown in Figure 5-8.

This also explains why one finds the nodules only at the top of the sediments. They seem almost to 'float' on the bottom of the ocean.

References

Berner, R. A.: 1964, *Geochim. Cosmochim. Acta* **28**, 1497.
Broecker, W. S., Gerard, R. D., Ewing, M., and Heezen, B. C.: 1961, *Am. Assoc. Advan. Sci. Publ.* **67**, 301.
Craig, H.: 1969, *J. Geophys. Res.* **74**, 5491.
Goldberg, E. D. and Arrhenius, G.: 1958, *Geochim. Cosmochim. Acta* **13**, 153.
Helgeson, H. C.: 1969, *Geochim. Cosmochim. Acta* **32**, 851.
Helgeson, H. C.: 1971, *Geochim. Cosmochim. Acta* **34**, 569.
Michard, G.: 1971, *J. Geophys. Res.* **76**, 2179.

CHAPTER 6

THE FRACTIONATION OF THE LIGHT ISOTOPES

Introduction

Modern mass spectrometers make it possible to measure variations in isotopic abundances to the order of 1 part in 10^4. With this sort of equipment, one might undertake to measure the isotopic composition of various elements in various natural mineral compounds, for example, the isotopic composition of S in natural sulfides, in native S, in S found in oils and coals, etc.

When this sort of systematic investigation is made for all the elements of the periodic table, it may be concluded that:

– For all elements of atomic number less than 20, there are isotopic variations detectable by modern means of measurement.

– For the elements with atomic number greater than 20, two cases must be distinguished:

(a) Most of the elements show a constant isotopic composition (the fluctuations are less than 1 in 10^4);

(b) Certain elements like Ar, Sr, and Pb show considerable variation in isotopic composition (sometimes by factors of 100 or 1000).

The collection of facts is interpreted as follows:

– The isotopic variations observed for the light elements are the result of isotopic fractionation associated with physical and chemical processes that are of the same type but much weaker than those that occur between two chemical elements.

These variations diminish as the atomic mass increases because the ratio $\Delta M/M$ diminishes (ΔM is the mass difference between isotopes of the same element), and once the atomic number 20 is exceeded, they have become so small that the present state of mass spectrometry can no longer detect them.

– The large variation in abundance of certain isotopes of certain elements like ^{87}Sr is the result of radioactive disintegration (thus the disintegration of ^{87}Rb produces ^{87}Sr).

Accordingly, the isotopic fractionation to be studied in this chapter takes place in physico-chemical processes associated with natural phenomena. The variations in composition that are caused by radioactivity will be studied in the last chapter.

6.1. The Evidence Concerning the Natural Fractionation of Isotopes

The systematic study of isotopic composition of the light elements in their various natural compounds discloses a certain number of systematic variations and of types of compositions.

As these variations in composition are extremely small, they are expressed in special units, the unit being δ, defined as:

$$\delta = \left(\frac{\text{isotopic ratio of the sample}}{\text{isotopic ratio of the standard}} - 1 \right) \times 1000.$$

The unit of δ is thus the relative deviation compared with the standard. The standards are chosen in such a way that the variations on all sides average out to $\delta = 0$, which then by definition is the standard.

Thus for O and H, the standard is mean sea water (SMOW) and for S is troilite from Canyon Diablo.

Figures 6-1 and 6-4 show the isotopic compositions of O and of S in the principal natural compounds.

Naturally, this first purely static approach has been followed by a series of studies seeking to explain these variations.

(1) Simple examination of the picture associated with the classical knowledge of geology and petrology allows correlation of the observed variations with definite geochemical processes. Thus for example, limestones precipitate from sea water by chemical and biochemical processes. Now it is reported that the isotopic composition of limestones in O is systematically high in ^{18}O. It is thus normal to suppose that

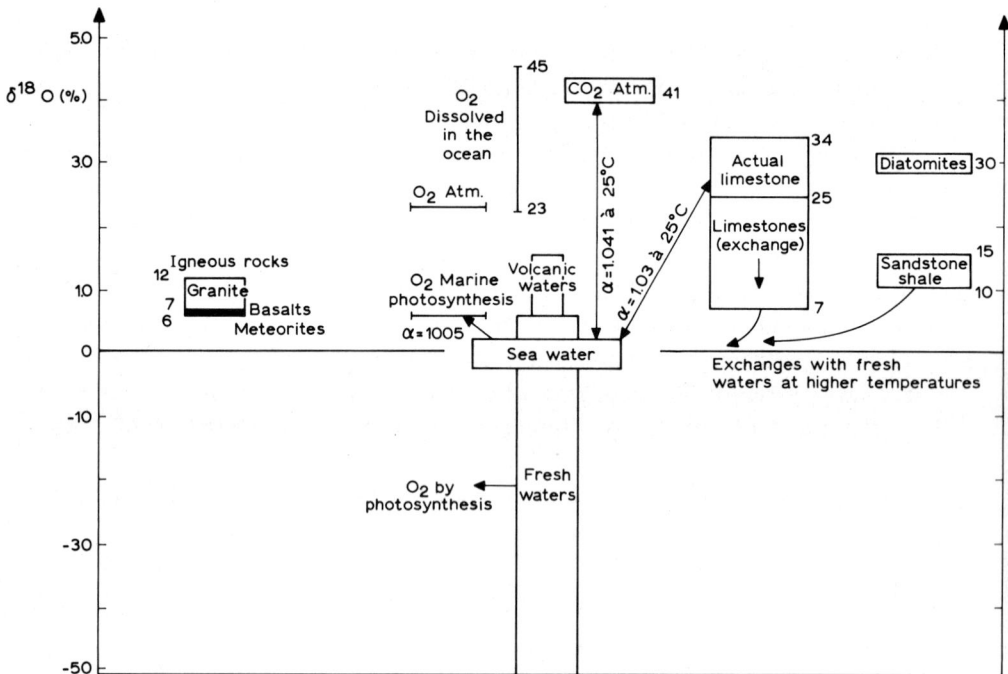

Fig. 6.1.   Variation of the isotope ratio for O (after Craig, 1963).

the formation of the limestones from sea water is accompanied by a certain isotopic effect.

In order to formulate the phenomenon quantitatively, a coefficient of global fractionation θ is defined.

Thus θ_{A-B} is the coefficient of fractionation between the two natural compounds A and B, and it is equal to the ratio of the isotopic compositions of A and B:

$$\theta_{A-B} = \frac{R_A}{R_B} \approx 1 + \frac{\delta_A - \delta_B}{1000}.$$

For the particular case of limestone in sea water:

$$\theta_{\text{limestone-sea water}} = \frac{(^{18}O/^{16}O)_{\text{limestone}}}{(^{18}O/^{16}O)_{\text{sea water}}} = 1 + \frac{\delta_{\text{limestone}}}{1000}$$

since by definition $\delta_{\text{sea water}} = 0$.

(2) From a physico-chemical point of view, studies have been made in the laboratory of various chemical reactions and various physical processes like diffusion or evaporation in order to determine the variation in isotopic composition that might be associated with these processes.

Thus for example, it has been reported that during the evaporation of water, the vapor becomes enriched in lighter isotopes, both for O and H. Careful measurements made in the laboratory have made it possible to define fractionation coefficients for each process. These basic fractionation coefficients are designated α.

Armed with these two types of information, geochemists have attempted to make a synthesis, that is to say, to tie together the θ's and the α's. Stated still differently, they have attempted to disentangle the natural phenomena into a series of elementary physico-chemical processes in such a way as to derive the natural coefficient θ.

This step is achieved practically by constructing physico-chemical models of natural processes and calculating the θ's from the α's measured in the laboratory. Once the agreement between the θ calculated and the θ observed in nature is satisfactory, it can be felt that the proposed model is a reasonable picture of reality. Thus by this approach, it becomes apparent that the study of the isotopic composition of natural compounds is not only interesting in its own right but in addition helps the understanding of the natural phenomena themselves. From this also emerges the use of radioactive tracers in geological phenomena for the light isotopes.

In groping to present the subject in a logical way, we shall ignore the historical development. We shall strive first to present the isotopic fractionations associated with various types of physico-chemical phenomena; then we shall explore some examples of isotopic fractionation in nature.

6.2. The Modes of Isotopic Fractionation

6.2.1. FRACTIONATION AT EQUILIBRIUM

As a result of the existence of several isotopes of a single element, the combinations among elements, that is, molecules and crystals, have many isotopic variations. Consider H_2O, for example. The following is the list of various isotopic combinations: $H_2{}^{18}O, H_2{}^{17}O, H_2{}^{16}O, D_2{}^{18}O, D_2{}^{17}O, D_2{}^{16}O, DH{}^{18}O, DH{}^{17}O, DH{}^{16}O$ (neglecting tritium, T). When the H_2O molecule is involved in an equilibrium process, all of these varieties are involved in the process, and the various equations of equilibrium must be written not just for the molecule H_2O but for each one of the corresponding isotopic molecules.

But still further, when two molecules do not react chemically or do not react further, or are in equilibrium and find themselves together, a new equilibrium called isotopic equilibrium is established between them.

Thus for example, the reaction:

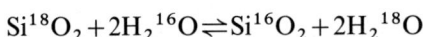

$$Si{}^{18}O_2 + 2H_2{}^{16}O \rightleftharpoons Si{}^{16}O_2 + 2H_2{}^{18}O$$

satisfies a law of mass action equation:

$$\frac{(H_2{}^{18}O)^2}{(H_2{}^{16}O)^2} \frac{(Si{}^{16}O_2)}{(Si{}^{18}O_2)} = K(T).$$

It is shown by statistical mechanics (Urey, 1947), and experiments confirm, that this constant is different from unity.

For an isotopic exchange reaction:

$$aA_1 + bB_2 \rightleftharpoons aA_2 + bB_1$$

B and A representing compounds, 1 and 2 indicating the existence of two isotopes of a common element in the two compounds.

We may write

$$K = \left(\frac{Q(A_2)}{Q(A_1)}\right)^a \left(\frac{Q(B_1)}{Q(B_2)}\right)^b.$$

The functions Q are called partition functions of the molecule and are such that for a given chemical species, one may write:

$$\frac{Q_2}{Q_1} = \frac{\sigma_1}{\sigma_2} \left(\frac{M_2}{M_1}\right)^{3/2} \frac{e^{-E_{2i}/kT}}{e^{-E_{1i}/kT}},$$

σ_1 and σ_2 are symmetry numbers of molecules 1 and 2. E_{2i} and E_{1i} are the different energy levels of rotation and vibration. M_1 and M_2 are the masses.

It thus appears that because of this mass factor, the larger the ratio M_1/M_2, the greater the degree of separation among the isotopes will be.

It further appears that $\log K$ can be expressed as $a' + b'/T + c'/T^2$, from which can be deduced that as $T \rightarrow \infty$, $K \rightarrow 1$. Thus at very high temperature, the isotopic fractionation approaches unity. If one defines, as was done above, an isotopic fractionation coefficient α associated with a process by the ratio:

$$\alpha = \frac{(A_1/A_2)}{(B_1/B_2)},$$

α and K are connected by the relation $\alpha = K^{1/n}$, n being the number of exchangeable atoms. In the preceding example, $n = 2$.

Such fractionation during processes of equilibrium is not limited to the case where the chemical species are different but is equally applicable during changes of phase. Thus during the evaporation of water, the vapor becomes enriched in the light isotope.

If the mixture of $H_2{}^{18}O$ and $H_2{}^{16}O$ is considered to be perfect and the water vapor an ideal gas, one may write:

$$pH_2{}^{16}O = xH_2{}^{16}O \cdot p^0(H_2{}^{16}O)$$
$$pH_2{}^{18}O = xH_2{}^{18}O \cdot p^0(H_2{}^{18}O)$$

x designating the molar fractions in the liquid and p^0 the saturation vapor pressure. p is the partial pressure of the designated species.

Then:

$$\alpha(\text{vap} - \text{liq}) = \frac{p^0(H_2{}^{18}O)}{p^0(H_2{}^{16}O)}.$$

The denser liquid being the less volatile, it follows that

$$p^0(H_2{}^{18}O) < p^0(H_2{}^{16}O) \quad \text{and} \quad \alpha < 1.$$

As for all fractionation coefficients, α depends on the temperature, and by use of the Clapeyron equation, it can be shown that $\log \alpha$ takes the form:

$$\log \alpha = \frac{a}{T} + b.$$

For water at $20\,^\circ\text{C}$,

$$\alpha(^{18}O) = 0.9911 \quad \text{and} \quad \alpha_D = 0.918.$$

At $20\,^\circ\text{C}$ the fractionation is around 8 times more important for D than it is for ^{18}O.

6.2.2. KINETIC FRACTIONATION

Molecules containing different isotopes do not react at identical rates. As a general rule, the lighter isotope reacts faster than the heavier one.

During a reaction, there will thus be a shift in composition between the starting substances and the end products.

For example, consider the reaction $C + O_2 \rightarrow CO_2$. Considering the oxygen isotopes there are two reactions:

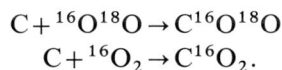

$$C + {}^{16}O^{18}O \rightarrow C^{16}O^{18}O$$
$$C + {}^{16}O_2 \rightarrow C^{16}O_2.$$

These two reactions will run with different speeds, with two kinetic rate constants, K_1 and K_2. We write u_1 and u_2 for the concentrations in the initial substances containing isotopes 1 and 2, and y_1 and y_2 for the concentrations in the end products. Then:

$$-\frac{du_1}{dt} = K_1 u_1 = \frac{dy_1}{dt}$$

$$-\frac{du_2}{dt} = K_2 u_2 = \frac{dy_2}{dt}.$$

If the concentrations of initial substances are held constant:

$$\frac{y_2}{y_1} = \frac{K_2 u_2}{K_1 u_1}.$$

One may also write: $\alpha' = K_2/K_1$.

This sort of fractionation occurs not only during chemical reactions but also during physical processes.

During evaporation of water, for example, carried out in a non-equilibrium process, that is, with the vapor being pumped away continuously, an isotopic fractionation is observed that proves to differ from the equilibrium value calculated just above.

Similarly, the processes of gaseous diffusion obeying Graham's law lead to a kinetic fractionation. For reactions involving gaseous oxygen, like those treated above, one has:

$$\frac{K_{16}}{K_{18}} = \sqrt{\frac{34}{32}} = 1.030 = \alpha'.$$

Oppositely to equilibrium fractionations that decrease with temperature, kinetic fractionation increases with temperature.

6.2.3. The transition of kinetic fractionation to equilibrium fractionation: isotopic exchange

Suppose two chemical compounds AO and BO are put in contact and that they have at least one element in common, for example, both having O in their formula. One of these species is assumed prepared with ^{18}O exclusively, the other with ^{16}O. At the end of a certain time in contact, it is supposed that the ratio $^{18}O/^{16}O$ of the two compounds is such that

$$\frac{(^{18}O/^{16}O)_{AO}}{(^{18}O/^{16}O)_{BO}} = K(T),$$

where $K(T)$ is the equilibrium constant.

Otherwise stated, the isotopes ^{18}O and ^{16}O have interchanged in such a way that equilibrium has been reached.

The reaction rate of this isotopic exchange can be measured, and it is found that:

(1) It is progressively faster as the temperature is raised.

(2) It is faster for gaseous or liquid media than if one of the compounds is solid. In the latter case, the rate of diffusion in the solid limits the kinetics of the process.

(3) It depends strongly on the steric locations of the O in the compounds AO and BO. The more nearly the O occupies an external location in the structure, the faster the kinetics of the process.

This isotopic exchange is of capital importance in geochemistry for it permits understanding of various basic observations.

(1) Let us suppose a reaction $A \to B$ occurs accompanied by a kinetic isotopic fractionation. If A and B are left in contact for a long enough time, the isotopes in A and B exchange until finally the distribution in A and B is of equilibrium type.

In order that the kinetic distribution be conserved, it is thus necessary to eliminate contact between the initial and final products.

(2) The exchange is favored by higher temperature. Thus at high temperature, the rapid and complete isolation of the product species prevents the attainment of equilibrium distribution.

6.3. Relationship between θ and α

Recollect that θ represents the observed fractionation whereas α represents the equilibrium distribution and α' the kinetic distribution.

We shall need to know whether the environment is open or closed and if the envisaged reaction leaves the initial and final products in contact or not. In the latter eventuality, one speaks of a blocked mechanism; after reaction, re-equilibration is no longer possible.

6.3.1. OPEN ENVIRONMENT

For a reaction with a non-blocked mechanism, global equilibrium is reached at $\theta = \alpha$.

With a blocked mechanism, $\theta = \alpha$ or α' depending on whether the distribution is an equilibrium one or a kinetic one.

6.3.2. CLOSED ENVIRONMENT

If the reaction mechanism is not blocked, there is again global equilibrium and $\theta = \alpha$, but the isotopic ratio for each term of the reaction varies so that the total quantity of each of the isotopes is conserved. In particular, if the reaction is complete, the reaction product has the same isotopic composition as the initial material prior to the start of the reaction.

If the mechanism is blocked, the partition coefficient applies only to the fraction of the product formed at the exact instant under study.

Effect of distillation. Let the phenomenon of transformation be $A \to B$, with which is associated for the element (i) a fraction characterized by the partition coefficient (α) defined for the two isotopes.

If the substance A is continuously renewed, the isotopic composition of product B (R_B) is equal to:

$$R_B = \alpha R_A.$$

If on the other hand, substance A is present in limited abundance, one can write:

$$\frac{dn_1/dn_2}{n_1/n_2} = \alpha,$$

where dn_1 is the amount of isotope 1 that is transferred from A to B, dn_2 is the amount of isotope 2 transferred from A to B.

On integrating, one obtains:

$$R_B = R_{0,A} f^{(\alpha-1)}$$

f being the fraction of A remaining, $R_{0,A}$ is the initial isotopic ratio of substance A; R_B is the isotopic composition of substance B found at the instant.

The mean composition of B is then:

$$\overline{R_B} = R_{0,A} \frac{(f^\alpha - 1)}{(f-1)}.$$

The process is one we have already encountered during the distribution of trace elements and is called Rayleigh distillation.

It plays a fundamental role in nature whenever a limited reservoir is involved.

Thus the evaporation of sea water obeys the equation:

$$R_B = \alpha' R_A$$

but the evaporation of a dead lake obeys the Rayleigh distillation law.

Having seen the broad general theoretical principles that regulate geochemical isotopes, we now turn to some geological applications.

6.4. The $^{18}O/^{16}O$ Isotopic Composition of the Silicates and Geothermometry

When the isotopic composition of O in silicate minerals is measured, one sees that it varies in a systematic way depending on the type of mineral and the type of rock to which the mineral belongs (Figure 6-1).

The composition can be characterized by measuring the isotopic fractionations among the various minerals.

Now one of the main interests in isotopes hinges on the fact that the isotopic fractions are independent of pressure, to a high degree of sensitivity. In effect, the changes in volume associated with exchange reactions are virtually zero.

The isotopic exchange equilibria can thus be highly useful for determining the temperatures of formation of mineralogical combinations. K varies with the temperature and approaches 1 at very high temperatures. The variation of K with T, as has been said, takes the form:

$$K = a_1 + \frac{b_1}{T} + \frac{c_1}{T^2}$$

and the form of this relationship is preserved with α and δ. Between two minerals A and B, in equilibrium:

$$\Delta_{AB} = \delta_A - \delta_B = a + \frac{b}{T} + \frac{c}{T^2} \simeq a + \frac{c}{T^2},$$

as the term in $1/T$ is generally negligible.

Oxygen isotopes are particularly useful for this purpose. Oxygen is the most abundant element in the silicates, and in addition the natural distribution of ^{18}O and ^{16}O occurs in fractions that are readily measurable mass spectrometrically.

The experimental studies of the Caltech group (Epstein and Taylon, 1967; Clayton and Epstein) give values of a and of c for a large number of mineral pairs (Figure 6-2).

If it can be assumed that isotopic equilibrium exists, then conversely, the measurement of Δ_{AB} in two minerals of a rock permits calculating its temperature of formation.

There is thus good reason to believe that equilibrium is established if all the mineral pairs in the rock lead to the same temperature. Javoy et al. (1970) have proposed a graphical method for this analysis. After selecting a reference mineral, one writes for each mineral

$$\Delta\,(\text{quartz, mineral}) - a = \frac{c}{T^2}.$$

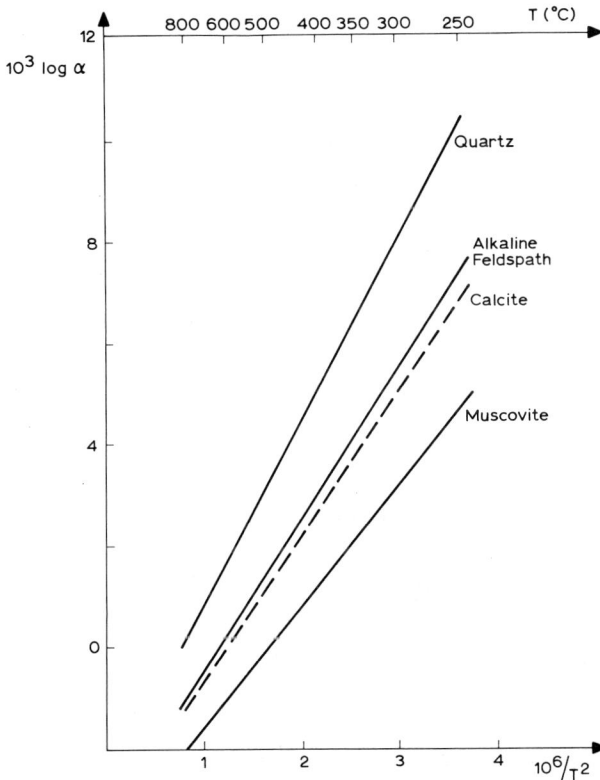

Fig. 6.2. Fractionation curves between water and several minerals as a function of T (after Taylor and Epstein, 1962).

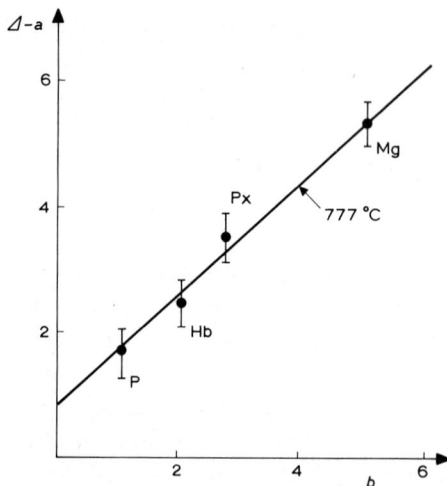

Fig. 6.3. Determination of the temperature of formation of the San Marcos gabbro.

In a graph of $\Delta - a$ as ordinate against c as abscissa, the different minerals of a rock in isotopic equilibrium yield points that fall on a straight line passing through the origin and with slope $1/T^2$, and this gives the temperature of formation. If the points fail to line up, the rock is not in equilibrium and the temperature of formation cannot be found (Figure 6-3).

It has thus become possible to paint a picture of the thermal domains in which the principal rocks were created. These results agree with the indirect evidence furnished by the experimental synthesis of minerals.

However, Taylor has recently shown that the interactions of silicates with water (whether magmatic, hydrothermal, or rain) may severely complicate the interpretation of the results in a large number of cases.

6.5. Isotopic Composition $^{32}S/^{34}S$ of Sulfide Ore Deposits

The majority of the deposits of economic interest of Pb, Zn, Cu, and Fe involve essentially sulfide minerals. For reasons that are simultaneously economic and geochemical, it is necessary to understand where the sulfides came from and how they accumulated.

If an inventory is made of the isotopic composition $^{32}S/^{34}S$ in natural sulfides as a function of their geological characteristics, one finds the following:

(1) All the sulfides in basic rocks have extremely constant compositions and close to the value $^{32}S/^{34}S = 22.22$ ($\delta = 0$), that is, approximately the same as for sulfides in meteorites.

(2) The sulfide minerals that are found in veins with gangues of quartz, fluorite, and barytine have values of $\delta \simeq 0$ that are very constant. It is therefore valid to suppose that they originate from a deep source.

(3) The minerals in lodes have much more variable compositions, in particular, the mineral accumulations found in sedimentary strata. The composition among these varies from $\delta = +22$ to $\delta = -52$.

This observation is associated with the fact that the oxidation-reduction reactions $S^{2-} \rightleftharpoons SO_4^{2-}$ are accompanied by isotopic equilibrium fractionation that at low temperature is considerable ($\alpha = 1.075$ at 25 °C). At low temperature, the oxidation reaction $S^{2-} \rightarrow SO_4$ is an easy reaction. On the contrary the reduction to SO_4^{2-} is possible only through the intervention of bacteria (*Desulfuvibrio desulfuricans*). This bacteriological reduction is accompanied by a slight isotopic effect, actually less than in the equilibrium reaction ($\alpha = 1.025$ at 25 °C). Remembering that the sulfates in sea water as well as in fresh water have δ values that range between $+26$ and $+4$, the observed dispersion might be explained by supposing that the sulfides bound in the strata come from the bacterial reduction of the sulfates but that this reduction offers variations. Sometimes it starts from sea water, sometimes from circulating subterranean water. Sometimes it takes place in a renewing environment, sometimes in closed reservoirs (Rayleigh distillation). Sometimes it is followed by an isotopic exchange leading to fractionation equilibrium, sometimes not. However, it is case by case, the isotopic composition of sulfur associated with metallogenic and geological observations that make it possible to limit the possible mechanisms admissible for the formation of the minerals.

In a more general way, these data have permitted establishing the presence of sulfide minerals of exogenic origin, a presence challenged up to the present by numerous authors, affirming that all mineralization comes from depths in the globe through the action of mineralizing fluids.

6.6. Some Aspects of the Water Cycle and the Associated Isotopic Fractionation

The water cycle of the Earth is dominated by the following factors:

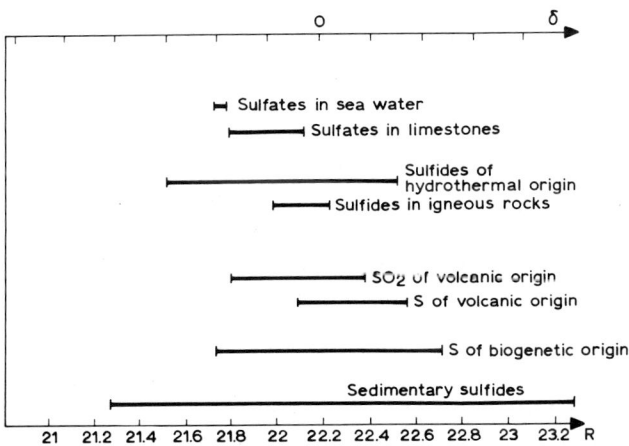

Fig. 6.4. Isotopic composition of S in some natural compounds.

(1) The cycle of evaporation and transport to clouds and precipitation constitutes the weather cycle;

(2) The existence of three great reservoirs, the oceans, the glacial caps, and the aggregation of fresh-water bodies, among which a series of exchanges occurs with or without the intervention of the meteorological cycle.

The totality of circulation of water on the globe and the diverse steps of the cycle have been studied from the isotopic point of view. We have seen during the theoretical examination that there were isotopic fractionations associated with the liquid-vapor equilibrium of water and that one could theoretically correlate the fractionations of O and H.

The use of this double isotopic pair has permitted the construction of quantitative models of the circulation of water. In any case, the problems posed by these studies are not so simple as those envisaged in the theoretical study. We shall see by way of example several aspects of these studies without much effort for their recent development is such that the aggregate of work is not yet sufficient to permit intensive study.

6.6.1. THE ISOTOPIC FUNCTIONING OF CLOUDS AND PRECIPITATION

Let us look at a cloud near the equator and follow its motion northward.

Step by step in the course of its evolution, the cloud discharges partially in the form of precipitation. Rainwater becomes enriched in heavy isotopes. In the process the cloud becomes enriched in light isotopes. Thus as time goes on, the precipitation material is richer in lighter isotopes, the effect being, however, partially compensated by the fact that the fractionation coefficient varies as $1/T$. In any case, as one goes from the equator northward, one finds statistically that the value of $\delta^{18}O$ grows more and more negative. It is the same with the composition of glaciers which only store precipitation water (Epstein, 1959). This general cycle of clouds repeats itself on a local scale when clouds penetrate into the continents and lose their water progressively. As a result, the fresh water has very large negative values of $\delta^{18}O$. This precipitated water has been studied with tracer pairs, $^{18}O/^{16}O$ and D/H, and it has been possible to show that precipitation water and the composition of glaciers fall on the same straight line (Figure 6-5).

The equation of the line is $\delta D = 8\delta^{18}O + 10$ in the diagram of δD vs $\delta^{18}O$. Now, we have seen that the slope 8 corresponds to the equilibrium fractionation value of water and its vapor. We are accordingly bound to believe that precipitation is a phenomenon that takes place under conditions of equilibrium (Figure 6-6).

It would seem on starting studies relating to the water cycle that similarly evaporation is statistically an equilibrium phenomenon. In fact, it is nothing of the sort. Evaporation is a kinetic type of phenomenon from the isotopic point of view, and the content of ^{18}O in the vapor is distinctly less than it would be at equilibrium. But, depending on the climate, kinetic evaporation is or is not followed by a partial isotopic re-equilibration that acts to prevent the vapor composition point of the graph from falling on the precipitation line. It is quite clearly the same with the surface layer of sea water which is the residue of evaporation.

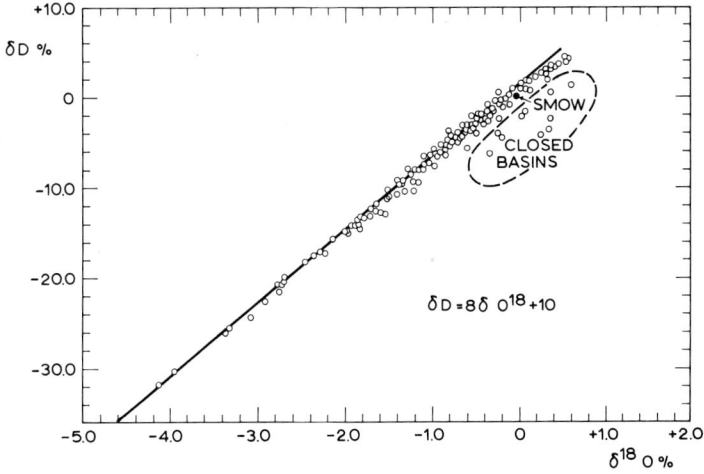

Fig. 6.5. The precipitation line (after Craig, 1963).

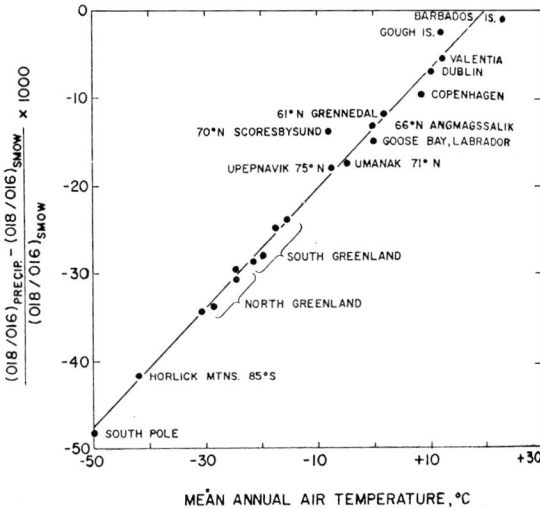

Fig. 6.6. Correlations between $\delta^{18}O$ vs Mean Air Temperature (after Dansgaard, 1964).

The content of ^{18}O in the surface sea water is variable and depends in fact on the relative importance of the evaporation of the precipitation that is very important over the ocean and the influx of fresh water (Figure 6-7). These variations are particularly noticeable in the North Atlantic (Epstein and Mayeda, 1953).

In order to illustrate these variations and to visualize the influence of the various phenomena that cause the variations, Craig (1965) proposed to plot the variations in a diagram of $\delta^{18}O$ vs salinity (expressed in parts per thousand).

6.6.2. THE GLACIERS, ISOTOPIC RECORD OF THE QUATERNARY CLIMATES

Another interesting application of this fractionation has been made by Epstein in the study of glaciers. When a core sample is taken from a glacier, the core shows stratified layers of ice that can be dated by various methods. The study of these layers of ice shows a variation in $\delta^{18}O$ and 8D (Figure 6-9a, b).

These variations for a given region are similar, and it becomes possible to find correspondence between the sequence pattern of one glacier with that of a neighboring one. One may thus hazard interpretations of these facts in two ways:

(a) Accepting that the origin of the precipitated water has varied during the course of recent geological time, this course gives a means of determining the fluctuations in the meteorological cycles of the past (Figure 6-8);

(b) Supposing that α has varied, then the observations disclose how the temperature has varied.

Recent studies show that in the Antarctic, it is the second effect that dominates, and it has been possible to determine the climatic variations of the Quaternary by this method. Remaining at present in dealing with mountain glaciers is the task of correcting for the effect of temperature, and of determining the paleo-circulation of clouds.

6.6.3. JUVENILE WATER

It is well known that in the water cycle, there is an addition from the hot waters rising from terrestrial depths. It was possible to believe for a long time that these hot waters sprang from the depths and were a progressive degassing of water trapped by the Earth at the time of its origin.

If this is really the picture, this water would progressively increase the volume of the hydrosphere. Water emerging in this way from the depths is called juvenile water.

Craig studied thermal spring water to determine the isotopic composition of this possibly juvenile water. He showed that in a graph of δD vs $\delta^{18}O$, the values for the thermal water from a single source fall on straight lines that are essentially parallel

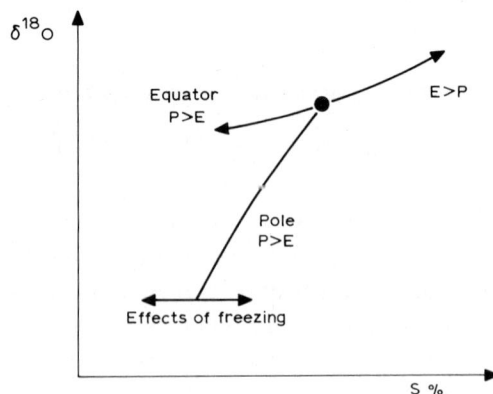

Fig. 6.7.   Evolution of $\delta^{18}O$ in sea water (after Craig, 1963).

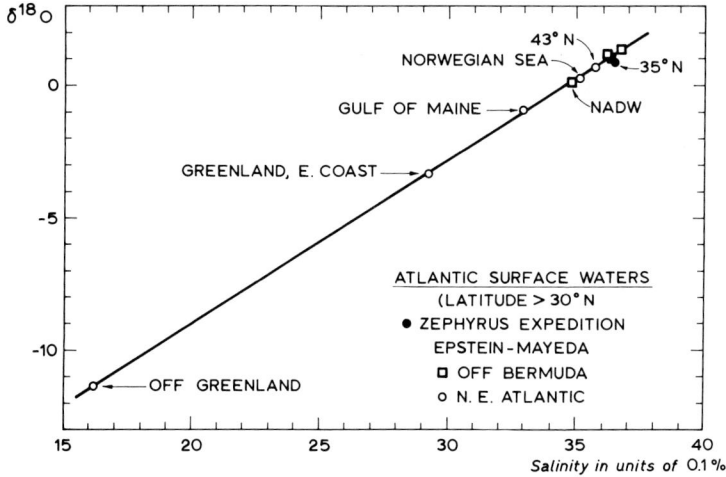

Fig. 6.8.   Graph of $\delta^{18}O$ vs S for the North Atlantic (after Craig, 1963).

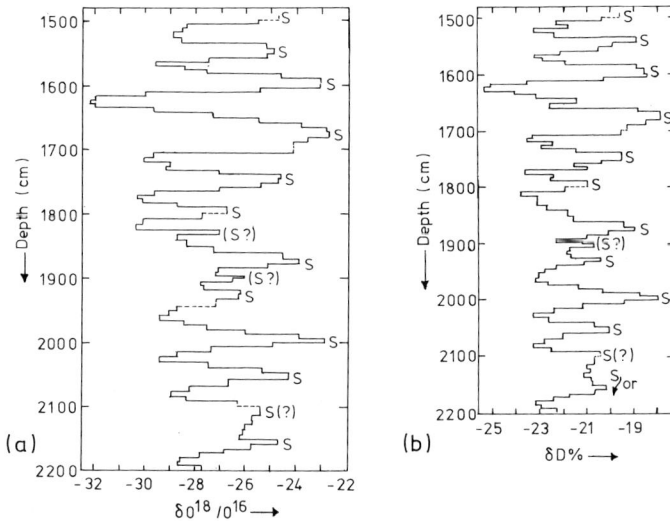

Figs. 6.9a–b.   Graphs of $\delta^{18}O$ and δD vs depth in the Antarctic Glacier. (a) Oxygen-isotope variations in Eights station firn core. S is a summer high; S (?), a questionable high. (b) Hydrogen-isotope variations in Eights station firn core. S is a summer high; S (?), a questionable high (after Epstein and Sharp, 1959).

to the $\delta^{18}O$ axis and cut the precipitation line at the point corresponding to the rain-water composition of the region. It thus becomes possible to explain the composition of thermal water by an evolution from meteorological water by O isotope exchange with the surrounding rocks. It is not necessary to call on juvenile water to explain the isotopic composition (Figure 6-10).

As these relations are systematic for all thermal regions studied, Craig concludes

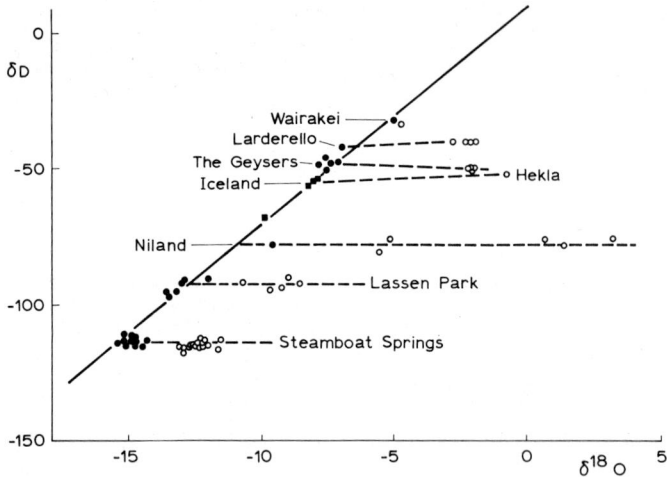

Fig. 6.10.   Isotopic composition of thermal water (after Craig, 1963).

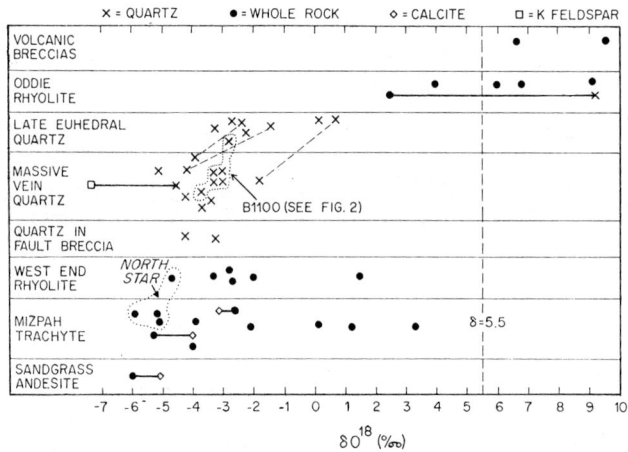

Fig. 6.11. $^{18}O/^{16}O$ Isotopic composition of mineral in rocks and minerals from the Tonopah district (after Taylor, 1973). Note the relative values to 5.5.

that the addition of juvenile water to the water cycle is negligible and that thermal springs contain only recycled surface water.

6.6.4. METEORITIC WATER – MAGMA INTERACTIONS IN THE GENESIS OF HYDROTHERMAL DEPOSITS

The study of stable isotopes, in particular O and H, led H. Taylor to the important discovery that unequivocal interactions existed between meteoritic water and volcanic – plutonic rocks. In his study of volcanism of the Scottish islands (Skye, Ardar-

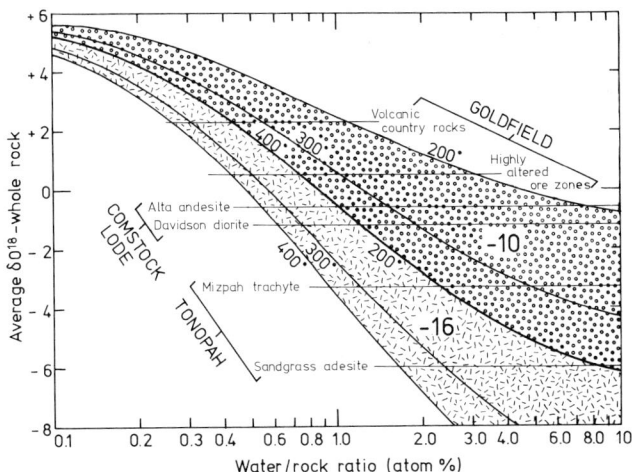

Fig. 6.12. Graph showing $^{18}O/^{16}O$ variation in exchangeable andrite with variable water/rock ratio.

muchan), he notes that:

- $\delta^{18}O$ values of total rocks may be abnormally low,
- Quartz – feldspars fractionation may be inverted.

Meteoritic waters are the only known source of low $\delta^{18}O$. On the contrary, juvenile waters have $\delta^{18}O = 5.5$ to 6.5 as shown by phlogopites of deep seated origin. Taylor's interpretation then is that meteoritic water exchanged with the volcanic rock while still hot. Since fractionation between minerals is low at high temperatures the net change is only a dilution effect which lowers the total rock $\delta^{18}O$. According to the variation of the water/rock ratio and temperature with time the phenomenon may become quite complex. For instance if the water/rock ratio is small the water isotopic composition will change toward higher $\delta^{18}O$. Such water allowed to exchange with a body of rocks at lower temperature may actually raise its $\delta^{18}O$ value. Taylor claims this to be operative in the Muskox intrusion to account for an increased $\delta^{18}O$ of the late terms. He further shows that $^{18}O/^{16}O$ and D/H systematics used complementarily favor this interpretation.

Using this theory Taylor showed that the mineralized quartz veins of the Tonopah district were produced by fluid of meteoritic origin which exchanged with andesitic and trachytic rocks at high temperature (Figure 6-11) forming one of the richest Au–Ag epithermal deposits (e.g. Comstock Lode and Goldfields, Nevada). He was also able to determine the water/rock ratio in this area (Figure 6-12).

Stable isotopes have provided an important tool in the study of mineral deposits and may help in their discovery.

References

Allègre, C. J.: 1966, *Mineralium Deposita* **1**, 104.
Bigeleisen, J.: 1949, *J. Chem. Phys.* **17**, 675.

Bottinga, Y.: 1969, *Geochim. Cosmochim. Acta* **33**, 49.

Craig, H.: 1963, in Tongiorgi (ed.), *Proc. Spolet. Conf. on Nuclear Geology*, p. 17.

Craig, H. and Gordon, L. J.: 1964, *Stable Isotopes in Oceanographic Studies and Paloetemperatures*, p. 9.

Epstein, S.: 1959, in P. Abelson (ed.), *Research in Geochemistry*, Vol. I, John Wiley, New York, p. 217.

Epstein, S. and Taylor, H. P.: 1967, in P. Abelson (ed.), *Research in Geochemistry*, Vol. II, John Wiley, New York, p. 29.

Javoy, M., Fourcade, S., and Allègre, C. J.: 1970, *Earth Planetary Sci. Letters* **10**, 12.

Taylor, H. P.: 1968, *Cont. Mineral. Petrol.* **19**, 1.

Taylor, H. P. and Epstein, S.: 1962, *Geol. Soc. Am. Bull.* **73**.

Thode, H. G.: 1963, in D. Shaw (ed.), *Studies in Analytical Chemistry*, Univ. of Toronto Press, Toronto.

Urey, H.: 1947, *J. Chem. Soc.*   , 562.

CHAPTER 7

FRACTIONATION OF RADIOACTIVE ISOTOPES AND ISOTOPES OF RADIOACTIVE ORIGIN

Introduction

As we have already emphasized, isotopic fractionation that accompanies physico-chemical phenomena is limited to the light elements ($M < 40$). However, it is established that for a certain number of elements of high atomic mass, there are great variations in the abundances of certain isotopes. For example, if the mass spectrum of one sample of Pb is measured that has been extracted from old zircon and of another sample from a galena vein (PbS), the spectra obtained are very different (Figure 7-1).

These considerable differences cannot be explained by physico-chemical isotope effects; they arise from a totally different cause – radioactivity.

It has been known since Aston's work that Pb has four isotopes, three are of radiogenic origin, that is, they originate partially from the radiodisintegration of two isotopes of U and one of Th. In this situation, furthermore, the respective decays from ^{238}U to ^{206}Pb, ^{235}U to ^{207}Pb, and ^{232}Th to ^{208}Pb are not direct but follow a series of cascading disintegrations (see Figure 7-2).

How then are the variations of mass spectra of natural leads to be explained from the above?

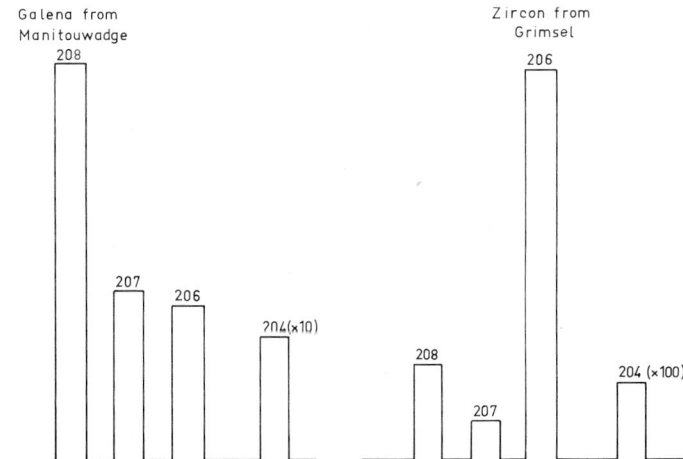

Fig. 7.1.   Isotopic composition of Pb from galena and zircon.

U	Pa	Th	Ac	Ra	Fr	Rn	As	Po	Bi	Pr	Tl
92	91	90	89	88	87	86	85	84	83	82	81

U_1 238 ----UX 234
$\;$ | UX₂‐UZ
U_2 234--- Io 230---Ra 226 ----Ra 222 --- Ra A 218 --- Ra B 214
$\qquad$ RaC 214⋯⋯RaC''210
$\qquad$ RaC' 214 -- RaD 210
$\qquad$ RaE 210
Th 232 ---MTh 1228 $\qquad$ Po 210 --- Pb 206
$\;$ | MTh 2228
RTh 228 --- ThX 224 ---- Tn 220 --- ThA 216 --- ThB 212
$\qquad$ ThC 212 ⋯⋯ ThC''208
UAc 235 -- UY 231 $\qquad$ ThC' 212 --- Pb 208
$\;$ | Pa 231 --- Ac 227 ⋯⋯ AcK 223
$\;$ RAc 227 --- AcX 223 --- An 219 --- AcA 215 --- AcB 211
$\qquad$ AcC 211 ---- AcC''
$\qquad$ AcC' 211 --- Pb 207

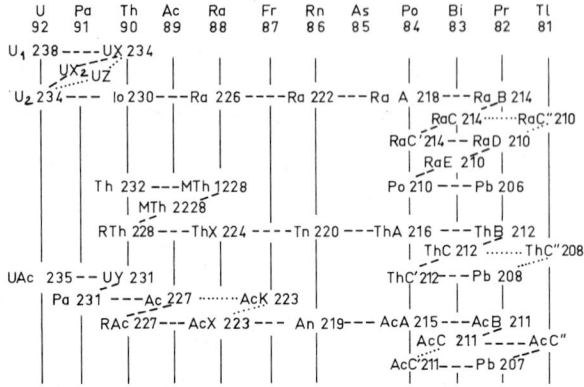

Fig. 7.2.　The natural radioactive series.

The law of radioactive distintegration, it may be remembered, takes the form:

$$\frac{dN}{dt} = -\lambda N,$$

where

N = number of parent atoms,
λ = radioactive decay constant,
t = time.

Integration then gives $N = N_0\, e^{-\lambda t}$. If M represents the number of daughter atoms produced by the disintegration, then

$$M = N_0 - N = N(e^{\lambda t} - 1),$$

where N_0 is the number of parent atoms at $t = 0$. Hence we may write, neglecting for the moment the occurrence of chain reactions:

$$^{206}\text{Pb}^* = {}^{238}\text{U}\left(e^{\lambda_1 t} - 1\right)$$
$$^{207}\text{Pb}^* = {}^{235}\text{U}\left(e^{\lambda_2 t} - 1\right)$$
$$^{208}\text{Pb}^* = {}^{232}\text{Th}\left(e^{\lambda_3 t} - 1\right).$$

Since Pb does possess a fourth isotope that is not of radioactive origin, ^{204}Pb and a part of the ^{206}Pb, ^{207}Pb and ^{208}Pb is also not radiogenic, we can thus write:

$$\frac{^{206}\text{Pb}_{\text{total}}}{^{204}\text{Pb}} = \frac{^{206}\text{Pb}^* + {}^{206}\text{Pb}_{\text{initial}}}{^{204}\text{Pb}}$$

$$= \left(e^{\lambda_1 t} - 1\right)\frac{^{238}\text{U}}{^{204}\text{Pb}} + \left(\frac{^{206}\text{Pb}}{^{204}\text{Pb}}\right)_{\text{initial}},$$

from which

$$^{(206/204)}\text{Pb}_{\text{actual}} = \frac{^{238}\text{U}}{^{204}\text{Pb}}\left(e^{\lambda_1 t} - 1\right) + {}^{(206/204)}\text{Pb}_{\text{initial}}$$

and also

$$(207/204)Pb_{actual} = \frac{^{235}U}{^{204}Pb}(e^{\lambda_2 t} - 1) + {}^{(207/204)}Pb_{initial}$$

and

$$(208/204)Pb_{actual} = \frac{^{232}Th}{^{204}Pb}(e^{\lambda_3 t} - 1) + {}^{(208/204)}Pb_{initial}.$$

It thus appears that the isotopic composition of a Pb sample will depend on the one hand on the initial isotopic ratios and on the other on the ratios $^{238}U/^{204}Pb$, $^{235}U/^{204}Pb$, and $^{232}Th/^{204}Pb$ in the environment in which the Pb 'evolved' in the course of history.

Returning to the opening example, Pb from zircon has evolved in an environment where $^{238}U/^{204}Pb \simeq 4000$, $^{235}U/^{204}Pb \simeq 29$, $^{232}Th/^{204}Pb \simeq 200$, with the initial ratios of $206/204 = 14.2$, $207/204 = 13.4$, $208/204 = 25$. Now galena evolved before separating out and becoming a mineral in a medium whose characteristics are $^{238}U/^{204}Pb \simeq 9.5$, $^{235}U/^{204}Pb \simeq 0.069$, $^{232}Th/^{204}Pb \simeq 2.7$ and the same initial ratios.

Suppose that the times of evolution are identical, for example, 2×10^9 yr, and that the constants, λ_1, λ_2, and λ_3 are those given in the table. The following isotopic compositions can be readily deduced (Table 7-I).

TABLE 7-I

Isotopic composition of lead from zircons and galenas

	$\lambda(10^{-9} \text{ yr}^{-1})$	Zircon	Galena
206/204	0.1537	1453	17.6
207/204	0.9722	187	13.8
208/204	0.0499	46	25.3

This very simple illustration nonetheless shows two highly important aspects of the phenomenon of radioactivity in geology:

(1) The isotopic fractionations of a substance are functions of the time; hence we have here a *method of dating*.

(2) The isotopic fractionations thus realized are tied directly to ratios like $^{238}U/^{204}Pb$. The isotopic ratios thus constitute a memory of the chemical history of the system in which these isotopic tracers evolved. We shall shortly see that the memory is much more subtle than might be believed from the foregoing example. In fact, we have treated a very simple example of the problem where the lead is in a chemically closed environment with constant chemical composition ($^{238}U/^{204}Pb = $ const.). We shall see that this is a very special case.

In nature, there exists a series of pairs (radioactive isotope – radiogenic isotope) for which the radioactive constants are small enough to lead to important natural fractionations. These have been assembled in the Table 7-II opposite. Having painted the picture of the subject with a broad brush, we shall return to draw it in finer detail.

TABLE 7-II

Principal mother-daughter pairs used in geochronology

Parent isotope	Daughter isotope	Decay constant	Remarks
^{87}Rb	^{87}Sr	1.39×10^{-11} yr^{-1}	
^{187}Re	^{187}Os	1.12×10^{-11} yr^{-1}	Difficult to use because Re/Os fractionation is very rare
^{40}K	^{40}Ca	4.18×10^{-11} yr^{-1}	Not usable because non-radiogenic ^{40}Ca is abundant
	^{40}Ar	5.88×10^{-11} yr^{-1}	
^{238}U	^{205}Pb	1.537×10^{-10} yr^{-1}	Radioactive decay series
^{235}U	^{207}Pb	9.72×10^{-10} yr^{-1}	Radioactive decay series
^{232}Th	^{208}Pb	4.99×10^{-11} yr^{-1}	Radioactive decay series

For all problems having to do with radioactive systems, the same concepts recur.

– The system of a parent and daughter isotope is assumed to be enclosed in a *box* at an *initial instant*. The box is defined by the spatial limits. A box may be a mineral: the initial time is then the epoch of formation of the rock. Or the box may be larger, for example, the planet Earth, or the Moon, or the upper mantle, or the sea, etc.

– Beside the radiogenic daughter isotope, and to serve as control, an isotope of the same element is used but one that is not radiogenic. That would be ^{204}Pb for lead and ^{86}Sr for strontium, ^{36}Ar for argon, etc.

– The box is then said to constitute a closed environment when neither parent nor daughter isotope, nor the reference isotope enters or leaves it.

In other cases, the box is an open environment. It is the latter case that is the more general in nature and we shall develop it a little.

7.1. General Equation for the Evolution of a Radioactive System (Wasserburg, Modified)

We consider the case of $^{87}Rb-^{87}Sr$, but the equations obtained apply equally well to other systems like $^{238}U-^{206}Pb$, $^{40}K-^{40}Ar$, etc.

It is assumed that the system $^{87}Rb-^{87}Sr$ is enclosed in a box that was created at the initial instant of our time scale.

This box exchanges matter with its external surroundings.

The equations describing the evolution of ^{87}Rb and of ^{87}Sr in the interior of the box are:

$$\frac{d^{87}Sr}{dt} = \lambda^{87}Rb - G(t) \cdot (^{87}Sr) + J(t),$$

where

$G(t)$ = the loss function of ^{87}Sr,
$J(t)$ = the flux of ^{87}Sr entering the system;

$$\frac{d^{87}Rb}{dt} = -\lambda^{87}Rb - H(t) \cdot (^{87}Rb) + U(t),$$

where

$H(t)$ = the loss function of ^{87}Rb,
$U(t)$ = the flux of ^{87}Rb entering the system.

As has been said, the evolution of ^{87}Sr is normalized in comparison with that of ^{86}Sr in order to interpret the evolution in terms of isotopic ratios.
For this non-radiogenic isotope, the equation may be written:

$$\frac{d^{86}Sr}{dt} = -G(t) \cdot (^{86}Sr) + J^*(t),$$

where

$J^*(t)$ = the flux of ^{86}Sr entering the system,
$G(t)$ = the loss function of ^{86}Sr, assumed to be equal to the loss function of ^{87}Sr.
It is possible to integrate this sytem of equations into the following form:

$$U(t) = \frac{^{87}Rb_{entering}}{^{87}Rb} \cdot {}^{87}Rb = V(t) \; {}^{87}Rb$$

and similarly

$$J(t) = \frac{^{86}Sr_{entering}}{^{86}Sr} \cdot {}^{86}Sr = L(t) \; {}^{86}Sr.$$

It then follows that

$$(^{87}Rb)_t = (^{87}Rb)_{t_0} \exp \langle \bar{K}(t) \rangle,$$

where

$$\langle \bar{K}(t) \rangle = \int_{t_0}^{t} [V(\tau) - H(\tau) - \lambda] \, d\tau$$

as well as

$$(^{86}Sr)_t = (^{86}Sr)_{t_0} \exp \langle \theta(t) \rangle,$$

where

$$\langle \theta(t) \rangle = \int_{t_0}^{t} [L(\tau) - G(\tau)] \, d\tau.$$

One may likewise write, noting that $^{87}Sr/^{86}Sr = \alpha$:

$$\left[\frac{d\alpha}{dt}\right]_t = \frac{d^{87}Sr}{dt} \cdot \frac{1}{^{86}Sr} - \frac{^{87}Sr}{^{86}Sr} \frac{d^{86}Sr}{dt} \frac{1}{^{86}Sr}$$

$$= \lambda \frac{^{87}Rb}{^{86}Sr} + \frac{J(t) - J^*(t)\left(\dfrac{^{87}Sr}{^{86}Sr}\right)}{^{86}Sr}.$$

It follows easily that:

$$\left[\frac{d\alpha}{dt}\right]_t = \lambda \frac{^{87}\text{Rb}}{^{86}\text{Sr}} + \left[\left(\frac{^{87}\text{Sr}}{^{86}\text{Sr}}\right)_{\text{ext}} - \alpha\right] L(t).$$

Noting that $(^{87}\text{Sr}/^{86}\text{Sr})_{\text{ext}} = \alpha^*$ and remembering that it is a function of t, we can write:

$$\left[\frac{d\alpha}{dt}\right]_t = \lambda \left(\frac{^{87}\text{Rb}}{^{86}\text{Sr}}\right)_{t=0} \exp[\langle \bar{K}(t) \rangle - \langle \theta(t) \rangle] + (\alpha^* - \alpha) L(t)$$

$$\times \left(\frac{^{87}\text{Rb}}{^{86}\text{Sr}}\right)_t = \left(\frac{^{87}\text{Rb}}{^{86}\text{Sr}}\right)_{t_0} \exp[\langle \bar{K}(t) \rangle - \langle \theta(t) \rangle].$$

If one admits that the box is surrounded by several systems with which it exchanges in the course of time, one may define for each system $(1, 2, ..., i, ..., n)$ the corresponding functions

$$G_1(t), G_2(t), ... \; G_i(t), ... \; G_n(t)$$
$$J_1(t), \; J_2(t), ... \; J_i(t), ... \; J_n(t)$$
$$H_1(t), H_2(t), ... \; H_i(t), ... \; H_n(t)$$
$$U_1(t), U_2(t), ... \; U_i(t), ... \; U_n(t),$$

one then defines:

$$\langle \bar{K}_{\Sigma}(t) \rangle = \int_{t_0}^{t} \left[\sum_{i=1}^{i=n} V_i(\tau) - \sum_{i=1}^{i=n} H_i(\tau) - \lambda\right] d\tau$$

$$\langle \bar{\theta}_{\Sigma}(t) \rangle = \int_{t_0}^{t} \left[\sum_{i=1}^{i=n} L_i(\tau) - \sum_{i=1}^{i=n} G_i(\tau)\right] d\tau.$$

The general equations then follow:

$$\left[\frac{d\alpha}{dt}\right]_t = \lambda \left(\frac{^{87}\text{Rb}}{^{86}\text{Sr}}\right)_{t=0} \exp[\langle \bar{K}_{\Sigma}(t) \rangle - \langle \bar{\theta}_{\Sigma}(t) \rangle] + \sum_{i=1}^{i=n} (\alpha_i^* - \alpha) L_i(t).$$

$$\left(\frac{^{87}\text{Rb}}{^{86}\text{Sr}}\right)_t = \left(\frac{^{87}\text{Rb}}{^{86}\text{Sr}}\right)_{t_0} \exp[\langle \bar{K}_{\Sigma}(t) \rangle - \langle \bar{\theta}_{\Sigma}(t) \rangle].$$

This equation is very general and is applicable to any box whatsoever, at any scale whatsoever, for instance, to one mineral, one rock, or to the upper mantle, or the ocean.

7.1.1. COMMENTS ON THE CONDITIONS OF INTEGRATION

As is usual in physics, the preceding equations are integrated with the assumption that $t=0$ at the time the system starts to function.

However, such a procedure leads to various origins for various events being studied.

By convention therefore, it is chosen to invert the time, and to consider that $T=0$ at the present time.

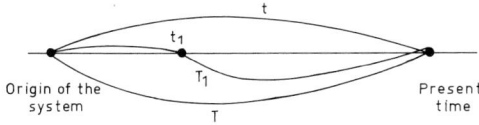

For greater convenience, the calculations are made with the usual conventions of physicists; then the origin is changed.

Special Cases

7.1.1.1. The System is Closed

In this case:

$$G(t)=J(t)=H(t)=U(t)=J^*(t)=0=(t).$$

It then follows simply that:

$$\frac{d\alpha}{dt}=\lambda\left(\frac{^{87}\text{Rb}}{^{86}\text{Sr}}\right)_o\exp(\lambda t)$$

$$\left(\frac{^{87}\text{Rb}}{^{86}\text{Sr}}\right)_t=\left(\frac{^{87}\text{Rb}}{^{86}\text{Sr}}\right)_{t=0}\exp(\lambda t).$$

The equation of the system then becomes at once:

$$\left(\frac{^{87}\text{Sr}}{^{86}\text{Sr}}\right)_t=\left(\frac{^{87}\text{Sr}}{^{86}\text{Sr}}\right)_{t=0}+\left(\frac{^{87}\text{Rb}}{^{86}\text{Sr}}\right)_t(e^{\lambda t}-1).$$

This is the classical chronometric equation for closed systems. The measurement of $(^{87}\text{Sr}/^{86}\text{Sr})$ on the one hand and of $(^{87}\text{Rb}/^{86}\text{Sr})_t$ on the other, and the knowledge of $(^{87}\text{Sr}/^{86}\text{Sr})_{t_0}$ permit calculation of the age t of the system. It should be noted that considering the present time as the origin does not change anything in the equations thus obtained.

7.1.1.2. The System Only Loses Elements without Gaining Any

This case can correspond to that of a mineral rich in Rb and Sr that tends to lose the excess of Rb and Sr to its surroundings in order to equalize the chemical potentials of the various minerals of the rock. This could also be the case for a permanent magmatic reservoir (portion of the upper mantle) that loses by volcanism a certain amount of Rb and Sr to the profit of the crust.

In this case, $J(t)=U(t)=J^*(t)=0$.

The general equation is identical, but the functions $\langle \bar{K}(t)\rangle$ and $\langle \theta(t)\rangle$ become simplified.

A considerably simpler case and one that at once permits treating several interesting problems is that for which:

$$G(t) = G = \text{constant with time}$$
$$H(t) = H = \text{constant.}$$

Then:

$$\alpha = \alpha_0 + \frac{\lambda \left(\dfrac{^{87}\text{Rb}}{^{86}\text{Sr}}\right)_{t=0}}{\lambda + H - G} \left(1 - \exp\left[-(\lambda + H - G)\,t\right]\right)$$

$$\left(\frac{^{87}\text{Rb}}{^{86}\text{Sr}}\right)_t = \left(\frac{^{87}\text{Rb}}{^{86}\text{Sr}}\right)_{t=0} \exp\left[-(\lambda + H - G)\,t\right],$$

with the origin of time at t_0. On shifting the time origin to the present, the equation remains unchanged.

7.1.1.3. *The Environment Gains Matter Constantly Without Losing Any*

At once, $G(t) = H(t) = 0$. This case corresponds to a mineral that is poor in Rb and Sr that gains these elements in the course of various processes. This could also be a reservoir that gains Rb and Sr continuously without losing any. In this case, the general equation is simplified through the simplification of the integrals $\langle \bar{K}(t) \rangle$ and $\langle \theta(t) \rangle$.

The simple cases can be interesting to examine, for example, by replacing $V(t)$ and $L(t)$ by constants and taking for α an exponential form (which returns to treating the surroundings of the system as an infinite medium with respect to the system). But these cases exceed the limits of this work for such models have not yet been the subjects of quantitative applications.

7.2. Evaluation of the Functions G(t) and J(t) and the Potassium-Argon Time Scale

Potassium 40 (^{40}K) disintegrates spontaneously giving ^{40}Ar and ^{40}Ca. In practice, only the method using $^{40}\text{K}/^{40}\text{Ar}$ is currently employed as ^{40}Ca is also the commonest isotope of Ca in nature. The chronometric equation is written:

$$\frac{^{40}\text{Ar}}{^{40}\text{K}} = \frac{\lambda e}{\lambda e + \lambda \beta} \left(e^{(\lambda e + \lambda \beta)\,t} - 1\right).$$

When the age of a mineral is measured by such a method, it is found that barring exceptions, complications intervene and in particular the chronometric findings obtained for various minerals from the same rock are different. These differences are attributed to the fact that Ar is highly mobile and that the systems do not remain closed.

In order to clarify these 'discordant ages', it has been attempted to measure the values of the functions $G(t)$ and $J(t)$.

7.2.1. DETERMINATION OF VALUES OF $G(t)$

It is here that the effects are the strongest for in general the ages determined from K–Ar are too young, which suggests that $G(t)$ dominates $J(t)$.

For this, one takes the situation where the time is short relative to the half-life of ^{40}K, and writes

$$\frac{d^{40}Ar}{dt} = -G(t)\,^{40}Ar.$$

$G(t)$ has the appearance of a kinetic constant and one can write:

$$G(t) = G_0 \exp\frac{-E}{RT(t)}$$

whence:

$$\log\frac{^{40}Ar}{^{40}Ar_0} = -\int_0^\tau G_0 \exp\frac{-E}{RT(t)}\,dt,$$

where

$^{40}Ar_0$ = quantity of Ar at the start of the experiment,

τ = duration of the experiment,

$T(t)$ = law of variation of temperature during the experiment, and

E = activation energy.

These studies have been done in two ways:

(a) In the laboratory, they have been done by subjecting the various minerals to progressive heating under various conditions (Figure 7-3). These experiments have enabled the determination of the G_0's and the activation energies (Figure 7-4).

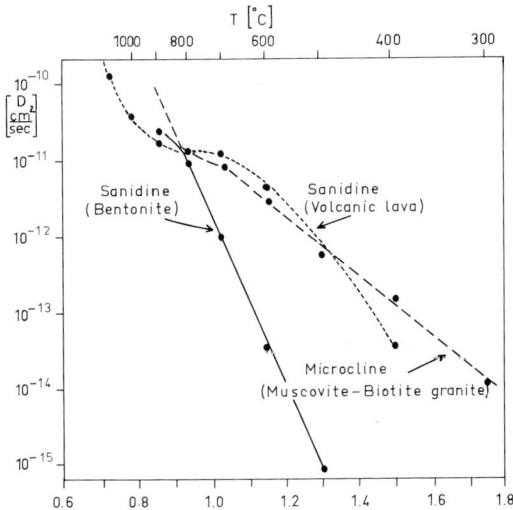

Fig. 7.3.   Diffusion coefficients for Ar in several minerals.

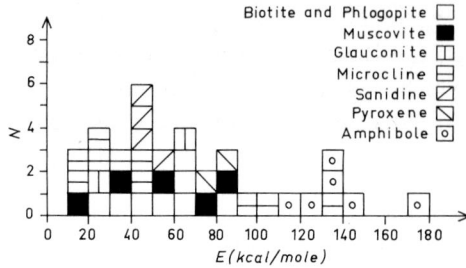

Fig. 7.4. Histogram showing activation energies for diffusion of Ar in several minerals.

(b) In the field, the studies have been done through recourse to the artifice of contact metamorphism (Hart, 1964). The intrusion of a recent granite into an old rock series provokes an intense reheating. The law of thermic evolution can be deduced if the petrographic nature and the geometry of the massif are known (Figure 7-5).

The values of the ratios of K to ^{40}Ar can be measured in various minerals at increasing distances from the contact.

The calculation then permits evaluation of the G_0's and the energy of activation for loss of Ar.

Although agreement is not very good quantitatively between the various field measurements and the various laboratory measurements, it has been possible to work out a scale of retentivities for the various minerals.

This scale shows that, in general, the age of hornblende < age of potash feldspar $\cong$ age of muscovite > age of biotite.

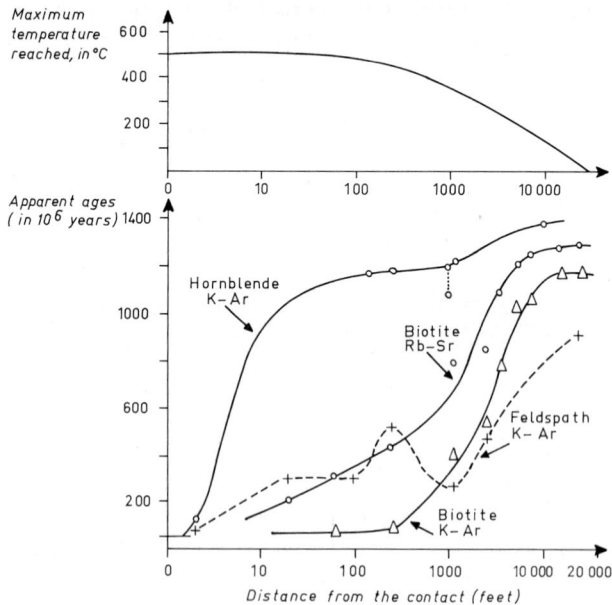

Fig. 7.5. Variations of apparent ages of minerals as a function of the distance from the contact surface in the case of contact metamorphism (after Hart, 1964).

By use of this scale, certain limits can be fixed, but alas, it is still difficult to get any farther (Damon).

7.2.2. EVALUATION OF $J(t)$

It is noted for certain minerals and in certain cases that the K–Ar ages are too old. This fact is interpreted as the result of an excess of Ar (Damon and Kulp, 1958).

These minerals are the beryls in pegmatites but also the pyroxenes and olivines in basic rocks and also listite in gneisses.

It is believed that during the genesis of the rock, the partial pressure of Ar in the vicinity of the mineral was high and the Ar became incorporated in the mineral.

In order to establish the value of the function $J(t)$ or more precisely $L(t)$, experiments at high pressure have been done to measure the solubility of argon in the minerals.

These experiments are unfortunately still in early stages, but they confirm that Ar is more soluble in the pyroxenes than in the calcic plagioclases.

All this work has as its goal the knowledge of the parameters that enter into the functions $G(t)$ and $J(t)$. It is to be hoped that once these parameters are known it will be possible in turn to calculate from measurements of $^{40}K–^{40}Ar$ on a series of minerals, the thermal history to which the rock has been subjected.

7.3. Isotopic Re-Equilibration of Strontium During Metamorphism

It has been shown above that the chronometric equation of a closed system for the pair $^{87}Rb–^{87}Sr$ had the form:

$$(^{87}Sr/^{86}Sr) = (^{87}Sr/^{86}Sr)_0 + (^{87}Rb/^{86}Sr)(e^{\lambda t} - 1).$$

Generally, the value of $(^{87}Sr/^{86}Sr)_0$ is not known, leaving us as a consequence with one measurement and two unknowns, t and the initial isotopic ratio of Sr.

In order to resolve the problem, it is possible to study a series of boxes of like genetic character for which an isotopic homogeneity can be assumed in advance to exist, permitting therefore the assumption that $(^{87}Sr/^{86}Sr)_0$ is identical in all the boxes.

In this case, a series of these similar boxes of the same age define a line in the graph of $^{87}Sr/^{86}Sr$ vs $^{87}Rb/^{86}Sr$ with slope $(e^{\lambda t} - 1)$ and an ordinate intercept of $(^{87}Sr/^{86}Sr)_0$ (Figure 7-6).

If the boxes are not cogenetic and have a variable ratio $(^{87}Sr/^{86}Sr)_0$, or if the system failed to stay closed, the experimental points would not fall on a line.

Thus when a series of cogenetic boxes is chosen to determine their age, it is essential not to stop with two boxes but to choose a much larger number in order to test the validity of the hypotheses used in setting up the basic equation (and in particular the hypothesis of a closed system).

In practice, work is carried out on two series of boxes. One consists of rocks from a single lithological suite – it may be said that one works on a system of whole rocks –, the other consists of the minerals in a single rock.

If, for a given granite massif, a study is made of its different mineralogical facies, as well as the minerals of each facies, two types of results are obtained (Figure 7-7).

(a) The line for the isotopic evolution of Sr defined by the analysis of the whole rock system crosses the lines for the isotopic evolution of the minerals.

The ensemble of systems, from the scale of the massif up to the scale of the minerals, remained a closed system after an episode of formation that isotopically homogenized the entire massif.

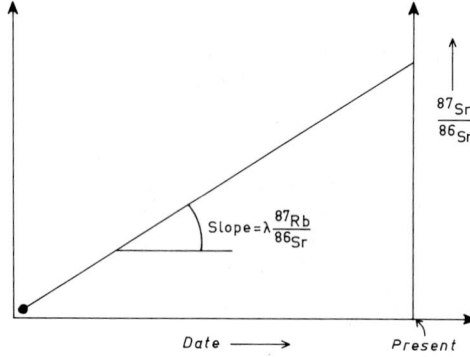

Fig. 7.6.   Idealized graph of $^{87}Sr/^{86}Sr$ vs $^{87}Rb/^{86}Sr$ for the case of a closed system.

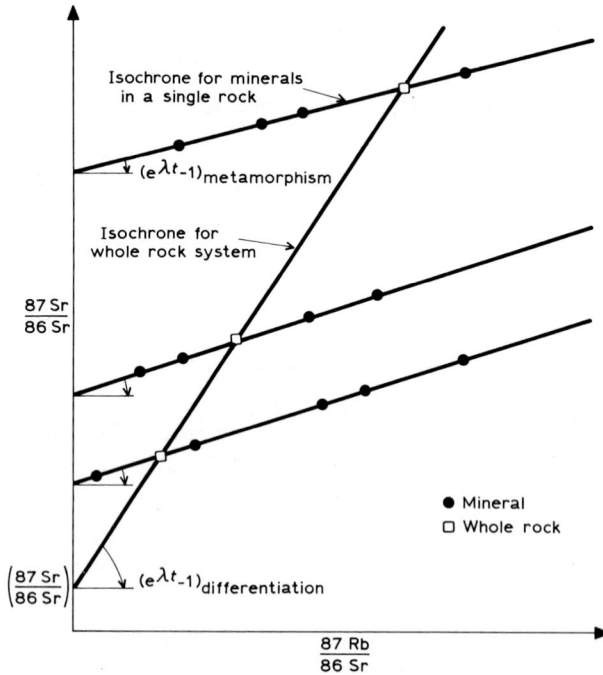

Fig. 7.7.   Rb-Sr diagram for a system in which the whole rock system remained closed but in which the minerals have been isotopically rehomogenized in the process of a metamorphism.

(b) The line for the isotopic evolution of Sr of the whole rock system has a greater slope (thus showing a greater age) than the line of isotopic evolution of the minerals. One is thus led to define two ages, one for the whole rock and one for the minerals.

It has been possible to show (Lanphere et al., 1964) that the age of the minerals corresponds to a later episode of metamorphism affecting the granites and that the whole rock age corresponds to the epoch of granite emplacement.

It seems therefore that a metamorphic episode stimulated a small scale migration of matter (^{87}Sr certainly, and possibly ^{87}Rb(?)) which in turn led to an isotopic re-homogenization of the minerals, but at the same time apparently did not affect the whole rocks, provided at least that the samples of these last were adequate in volume.

Recent studies carried out on this problem have shown that starting from these two cases, there may prove to be other variations.

(a) When the perturbing metamorphism is mild, the exchange of matter among minerals is only partial, and the isotopic homogenization is also only partial. The points representing the various minerals thus form a 'cloud' in the graph of isotopic evolution. In this exchange one can identify minerals that are rich in ^{87}Sr of radiogenic origin and hence have a tendency to lose it (for example, biotite). These are donor minerals. Those that are poor in Rb (hence also in radiogenic ^{87}Sr) but rich in non-radiogenic Sr and hence have a tendency to capture the radiogenic ^{87}Sr are the acceptor minerals.

It is known that radiogenic ^{87}Sr is the element that migrates most readily. Since it arises from the disintegration of ^{87}Rb, it occupies the crystallographic sites of Rb in which it is 'ill at ease'. It tends therefore to migrate to sites that are more suitable to it.

If one supposes that only the ^{87}Sr is mobile, which is reasonable as a first approximation, and that the period of metamorphism was reasonably brief, the general equation may be written in a slightly modified form.

For the donors, and during the metamorphic event:

$$\frac{d^{87}Sr}{dt} = -G(t)\,(^{87}Sr),$$

^{87}Rb and ^{87}Sr being constants.

For the acceptors,

$$\frac{d^{87}Sr}{dt} = J(t) = J_1(t) + J_2(t) + \dots,$$

where $J_1(t)$, $J_2(t)$ are the fluxes from the donor minerals 1, 2,....

An interesting case is that of donor-acceptor pairs for which everything lost by the donor is taken up by the acceptor (which, furthermore, receive only from this one donor).

This is the case for example, of the association of biotite and apatite at the onset of metamorphism or potash feldspar with plagioclase when the metamorphism is a little more intense.

If it is remembered that L is very large, it can be seen that at the start of the process, the ratio $(^{87}Sr/^{86}Sr)$ for the acceptor (which is much less abundant than the donor in the case of the apatite-biotite pair) changes much more than the ratio $(^{87}Sr/^{86}Sr)$ of the donor.

Measurement of the ratio $(^{87}Sr/^{86}Sr)$ for minerals like apatite or epidote thus serves as a means of detecting the movement of matter even when the metamorphic event is still very weak (Figure 7-8).

This has been excellently shown for the first time by Wasserburg and Steiger (1967).

It should be noted that from the purely chronological point of view, if one takes as the initial ratio of $^{87}Sr/^{86}Sr$ the whole rock values, the donors apparently have a lesser age than initial age (rejuvenation). The acceptors appear older by comparison with the initial age.

(b) Another extreme case is that for which the metamorphism is sufficiently intense to cause migrations across the boundary of the whole rock system. The phenomenon of deep anatexis is the most significant example. When it occurs, it re-homogenizes the rocks isotopically over a distance of several kilometers.

By contrast, in the granulite facies, that correspond however to higher (P, T) conditions than for the anatexis zone, the isotopic homogenization is only very partial and it is possible to discover the age of the formation from the isotopic line of the graph. This difference of behavior of rock systems as to isotopic homogenization of Sr is to be regarded as parallel with the existence of freely circulating water in the anatexis zone and the absence of free water in granulite facies.

On a lesser scale, and under less intense conditions of metamorphism, migrations occur in the course of metamorphism when the chemical gradients of radiogenic ^{87}Sr are significant. For example, when a contact between basic rock and acid rock is subjected to metamorphism, the ^{87}Sr migrates from the acid rock toward the basic

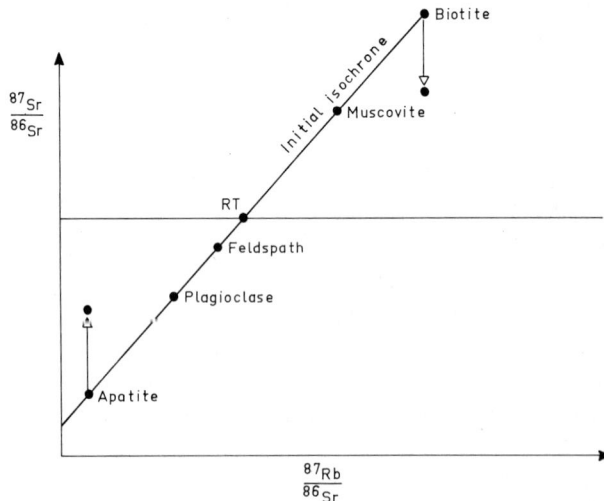

Fig. 7.8.   Rb-Sr diagram for a system of cogenetic minerals showing the evolution of the donor-acceptor pair biotite-apatite.

rock, producing in the latter absurd apparent ages (Wasserburg, *et al.*, 1964). These migrations do not exceed a few meters.

We have covered in some detail the case of the system ^{87}Rb–^{87}Sr, but it is not necessary to confine the conclusions that are reached to this single example.

The systems U–Th–Pb are susceptible to similar reasoning, but the donor minerals are zircon, sphene, or apatite while the acceptors are largely K feldspars.

Similarly, the K–Ar system also behaves in an analogous way, and it is not necessary to believe as is generally done that only the losses of Ar are to be taken into account in this type of chronology. The gains in Ar (one commonly says the excess of Ar) are quite frequent, even in the potassic minerals like biotite (Damon, Giletti).

7.4. Loss of Radiogenic Elements in Rich Systems and Chronology of U–Th–Pb From Zircons

Zircons are the common minerals by far the richest in U and Th. They are so rich in these elements that the radiogenic component of Pb completely masks the initial component of Pb_1. As a consequence in correcting the initial Pb content in an arbitrary way one makes only a most trifling error in the case of the zircons. It is therefore possible to consider the system U–Th–Pb in zircons as exclusively radiogenic.

The chronometric equations of a closed system take the form in this case:

$$\frac{^{206}\text{Pb}}{^{238}\text{U}} = (e^{\lambda t} - 1)$$

$$\frac{^{207}\text{Pb}}{^{235}\text{U}} = (e^{\lambda' t} - 1)$$

$$\frac{^{208}\text{Pb}}{^{232}\text{Th}} = (e^{\lambda'' t} - 1).$$

A test of the validity of these equations can be gotten here in a simple way. It is, in effect, possible to calculate an age for a single sample of zircon in three different ways. If the system has remained closed, these three ages prove to be identical. This situation arises in fact in a certain number of cases and more frequently still for other radiogenic minerals, namely, the sphenes (Tilton and Grünfelder, 1960). It has proved possible to confirm this age with measurements on Rb/Sr.

Unfortunately, in the great majority of cases, the ages obtained by the closed model do not agree. A logical hypothesis is then to admit that the system does not behave like a closed but like an open system. Since it is a system that is very rich in U, Th, and radiogenic Pb, one may suppose that the influx transport functions are zero and only the loss functions come into play.

For a given system, limited in consideration to the portion of Pb that is radiogenic, the equations become, for example:

$$\frac{d^{206}Pb}{dt} = \lambda^{238}U - G(t)(^{206}Pb)$$

$$\frac{d^{238}U}{dt} = -\lambda^{238}U - H(t)(^{238}U).$$

Calling r_λ the ratio $^{206}Pb/^{238}U$, it follows that:

$$\frac{dr_\lambda}{dt} = \lambda + r_\lambda[H(t) - G(t) + \lambda].$$

One may write the same for $r_{\lambda'} = {}^{207}Pb/^{235}U$ and $r_{\lambda''} = {}^{208}Pb/^{232}Th$:

$$\frac{dr_{\lambda'}}{dt} = \lambda' + r_{\lambda'}[H'(t) - G'(t) + \lambda']$$

$$\frac{dr_{\lambda''}}{dt} = \lambda'' + r_{\lambda''}[H''(t) - G''(t) + \lambda''].$$

It is clear that a measurement of U–Th–Pb on a specimen of zircon does not permit solving the system of proposed equations without a precise knowledge of the functions $G(t)$, $H(t)$, $G'(t)$, $H'(t)$, $G''(t)$, and $H''(t)$.

A simplification is always possible if it is noticed that the two systems involving U–Pb are identical from the point of view of chemistry. It is believable that the loss functions of the two U–Pb systems are identical, hence that:

$$G(t) = G'(t); \qquad H(t) = H'(t).$$

This offers an invitation therefore to compare the results, not from a specimen of zircon, but from a population of zircons, supposedly of the same age (for example, extracts from a single rock).

This examination can be made graphically in the following way:

Consider a graph of $^{207}Pb/^{235}U$ as abscissa and $^{206}Pb/^{238}U$ as ordinate. Following Wetherill (1956) a curve is defined corresponding to closed systems (Figure 7-9). It is mathematically defined by the equations $r_\lambda = (e^{\lambda t} - 1)$, $r_{\lambda'} = (e^{\lambda' t} - 1)$.

The curve, again following Wetherill, is named the Concordia curve. Each specimen of zircon that gives the same age by both clocks, $(^{238}U/^{206}Pb)$ and $(^{235}U/^{207}Pb)$, gives a point on the Concordia curve. Each sample that gives different ages by the two methods (so-called discordant ages) lies off of the Concordia curve. If various specimens of zircons are separated belonging to one and the same rock according to various criteria, size and magnetic susceptibility, and if the ratios r_λ and $r_{\lambda'}$ are measured, their graph on the Concordia plot shows them on a straight line that cuts the Concordia line in two points (Figure 7-9).

These two intersections correspond to two concordant ages that we designate t and t_1.

Wetherill sought to interpret this regularity, first shown to exist by Ahrens (1955),

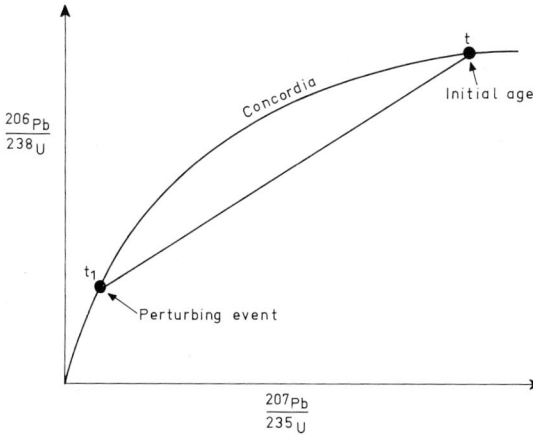

Fig. 7.9.   Concordia diagram showing Wetherill's model (1956).

in the following way: He assumed that $G(t)$ and $H(t)$ must be discontinuous functions of time and that the losses occurred exclusively during a metamorphic event.

We write:

$$H(t) - G(t) = a_i \delta(t - t_i),$$

[where $\delta(t - t_i)$ is the Dirac delta function with value zero everywhere except $t = t_i$ and $\int \delta(t - t_i)\, dt = 1$].

Integration is carried out by taking the present time as origin. The result is:

$$r_\lambda = R\left[e^{\lambda t} - e^{\lambda t_1}\right] + \left[e^{\lambda t_1} - 1\right],$$

where $R = e^{a_i}$ is the loss coefficient during the metamorphic event.

One may also write:

$$r_{\lambda'} = R\left[e^{\lambda' t} - e^{\lambda' t_1}\right] + \left[e^{\lambda' t_1} - 1\right].$$

The two equations for a series of zircons of like age t and having experienced the same metamorphic event at t_1 (but with different intensities) then become:

$$\frac{r_\lambda - (e^{\lambda t} - 1)}{r_{\lambda'} - (e^{\lambda' t} - 1)} = \frac{e^{\lambda t} - e^{\lambda t_1}}{e^{\lambda' t} - e^{\lambda' t_1}}.$$

They thus define a line on a graph of r_λ vs $r_{\lambda'}$ that cuts Concordia at the two points corresponding to t and t_1.

The interpretation of t and t_1 given by Wetherill is thus that these points correspond respectively to the initial age of the population and to the age of the metamorphic event. This interpretation is entirely in accord with the experimental results of Silver (1963) (Figure 7-10).

It is of some interest to emphasize that by entirely different routes, the method of the isotope line on the one hand and the Concordia graph on the other leads to

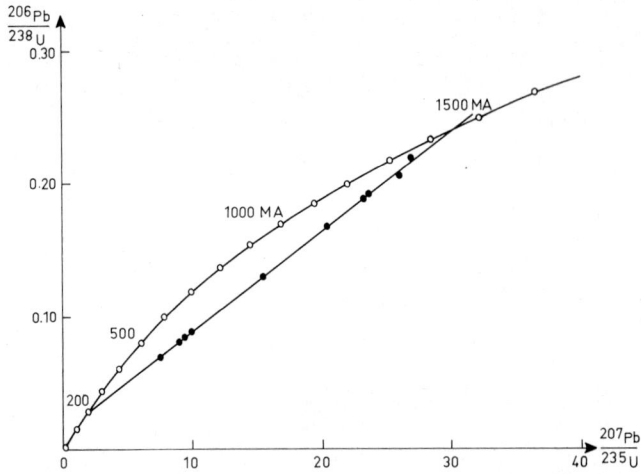

Fig. 7.10.   Concordia diagram of a family of cogenetic zircons (after Silver, 1963).

determinations of initial age of formation of a rock and the age of a metamorphic event.

Following Tilton (1960) and later Ulrych (1963), another interpretation is possible for the existence of the alignment on the Concordia graph. Suppose that instead of losing the Pb in a discontinuous way in the course of a metamorphic event, the system loses the Pb continuously.

The equations then become (actually, Tilton used a diffusion equation that is a little different from ours here):

$$\frac{dr_\lambda}{dt} = \lambda + r_\lambda[G + \lambda]$$

assuming that the loss continues at a constant rate. Integration then gives

$$r_\lambda = \frac{\lambda}{\lambda - G}[e^{(\lambda - G)t} - 1]$$

$$r_{\lambda'} = \frac{\lambda'}{\lambda' - G}[e^{(\lambda' - G)t} - 1]$$

with the time origin being taken as the actual time.

These parametric equations represent the evolution of a population of zircons of the same age losing Pb by a continuous process with various intensities of the loss G. The curve representing such equations on the Concordia graph is a straight line in its first portion.

Tilton deduces therefrom that the alignment observed in the diagram proves nothing concerning the existence of a metamorphic event but fits well into the framework of continuous loss. In this situation, the lower intersection of the line with the Concordia curve has no geological significance (Figure 7-11).

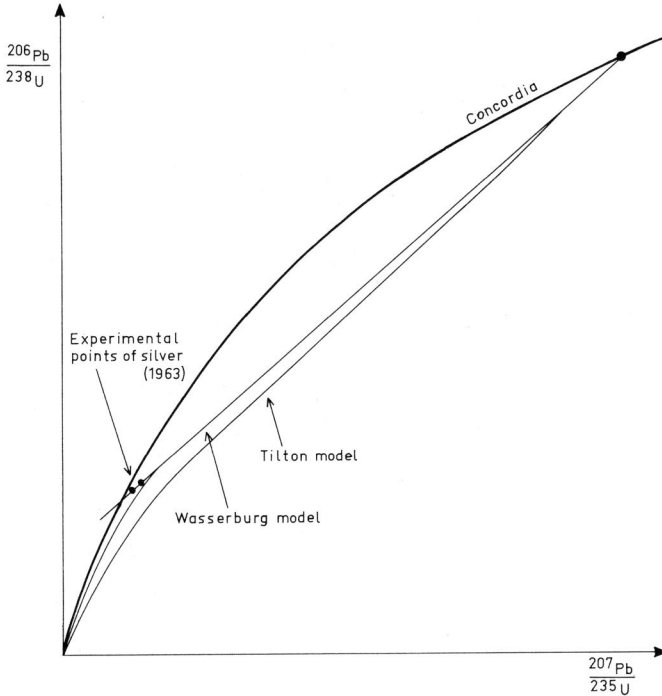

Fig. 7.11.   Concordia diagram showing Tilton's interpretation. The model developed by Wasserburg (1963) is also shown as well as the experimental objections of Silver.

Without entering into the developmental details (Wasserburg, 1963; Silver, 1963) or discussions that followed the proposing of the two fundamental and contradictory interpretations, a certain number of points of the method may be emphasized.

In the example of the isotope evolution line, use was made of a chronometer and a series of different, cogenetic boxes (minerals and whole rocks). Here by contrast we use the same boxes but two different chronometers that we have grouped together by virtue of their similarities in behavior.

This idea of using two coupled systems for interpreting the ages of a single system of boxes with a different constant is extended to the U–Pb and Th–Pb systems as well as to the K–Ar and Rb–Sr systems or the Io–U and Pa–U (Allègre, 1964) systems and thus seems to be a highly general method for radiometric age determinations.

We note in closing that the information obtained by such a type of analysis is not limited to dating but covers also the chemical fractionation (U/Pb for instance) that rock systems have undergone in the course of geological phenomena.

7.5. Evolution and Nature of the Earth's Mantle

We shall now change scales and attempt to study the evolution, no longer of systems of minerals, but of much more important geological units, for example, the crust or the terrestrial mantle. We shall start with the latter.

As has been brought out, the origin of basaltic magmas must be explored in the upper mantle. If the rocks are considered for which the extent of partial fusion is adequate and which were created outside of continental zones, it is believable that the isotopic composition of Sr or of Pb in these rocks is the same as in the mantle zones from which they came.

Let us examine the isotopic composition of Sr in oceanic basalts (this composition is measured by the ratio $^{87}Sr/^{86}Sr$).

Tholeites yield a uniform composition equal to 0.7027 (Hart, 1964). Alkaline basalts have different compositions that vary between 0.703 and 0.705 (Hedge and Peterman, 1970). The results make it possible to establish limits on the nature and evolution of the terrestrial mantle.

Model 1 (Figure 7-12). Assume that the mantle remained a closed environment since the origin of the Earth. Then one may write:

$$\left(\frac{^{87}Sr}{^{86}Sr}\right)_{\substack{present \\ basalts}} = \left(\frac{^{87}Sr}{^{86}Sr}\right)_{\substack{at\ the\ origin \\ of\ the\ Earth}} + \left(\frac{^{87}Rb}{^{86}Sr}\right)_{mantle} (e^{\lambda T_0} - 1).$$

The values of the constants are known:

$$T_0 = 4.6 \times 10^9 \text{ yr}$$

$$\left(\frac{^{87}Sr}{^{86}Sr}\right)_{\substack{at\ the\ origin \\ of\ the\ Earth}} = 0.6989.$$

Measurement of the ratio $(^{87}Sr/^{86}Sr)_{recent\ basalts}$ then furnishes a value for the ratio $(^{87}Rb/^{86}Sr)$ in the mantle.

The values thus obtained permit the conclusion that within the framework of this model:

(1) The mantle is chemically homogeneous.

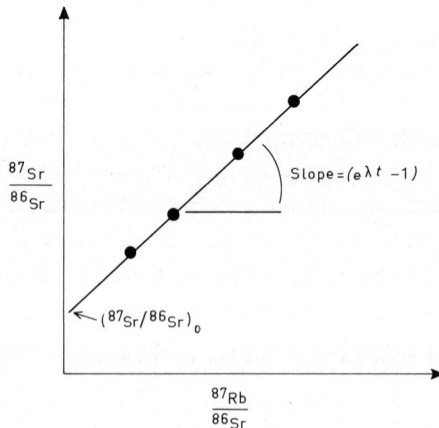

Fig. 7.12.   Graphical form of the equation for the evolution $^{87}Sr/^{86}Sr$ as a function of time (closed environment).

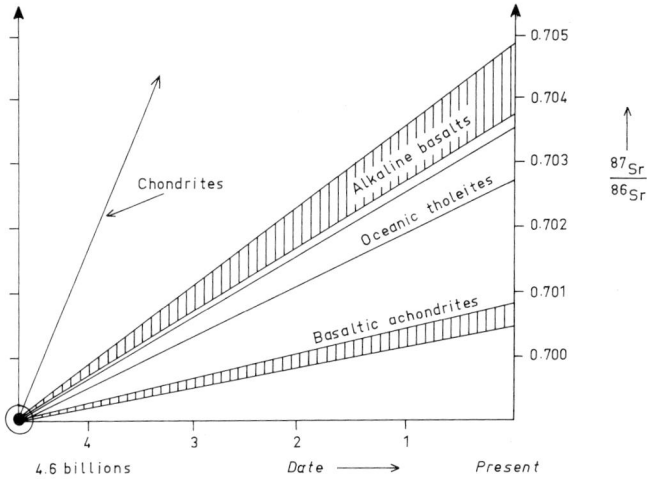

Fig. 7.13.   Graph of $^{87}Sr/^{86}Sr$ vs time showing the evolution of various rocks supposing that they evolved in a closed environment.

(2) The values of the ratio $^{87}Rb/^{86}Sr$ calculated for the mantle are distinctly less than those for chondrites and much nearer to the values for basaltic achondrites (Gast, 1958) (Figure 7-13).

Model 2 (Figure 7-14). The mantle was originally homogeneous, but it suffered a chemical fractionation of Rb/Sr such that the mantle can be assumed to have evolved in an open environment.

These fractionations might be volcanic phenomena that are known to enrich the melt in Rb by comparison with Sr. Thus domains differentiate in the mantle in which the ratio of Rb/Sr varies with time.

This could happen in a continuous or a discontinuous manner.

We turn next to the isotopes of Pb (Tatsumoto, 1966).

These isotopes have the advantage of possessing two isotopic pairs:

$$^{206}Pb/^{204}Pb = \left(\frac{^{206}Pb}{^{204}Pb}\right)_0 + \frac{^{238}U}{^{204}Pb}(e^{\lambda t_0} - 1)$$

$$^{207}Pb/^{204}Pb = \left(\frac{^{207}Pb}{^{204}Pb}\right)_0 + \frac{^{235}U}{^{204}Pb}(e^{\lambda' t_0} - 1)$$

with $^{238}U/^{235}U = 137.8$; $^{206}Pb/^{204}Pb = 9, \ldots$; $^{207}Pb/^{204}Pb = 10, \ldots$.

These equations may be plotted in a graph of $^{206}Pb/^{204}Pb$ vs $^{207}Pb/^{204}Pb$ (Gerling plot).

In such a graph, all the leads that have evolved up to the present in a closed environment fall on a line characterized by the equation (Figure 7-15).

$$\frac{\left(\frac{^{206}Pb}{^{204}Pb}\right) - \left(\frac{^{206}Pb}{^{204}Pb}\right)_0}{\left(\frac{^{207}Pb}{^{204}Pb}\right) - \left(\frac{^{207}Pb}{^{204}Pb}\right)_0} = 137.8 \frac{(e^{\lambda T_0} - 1)}{(e^{\lambda' T_0} - 1)}.$$

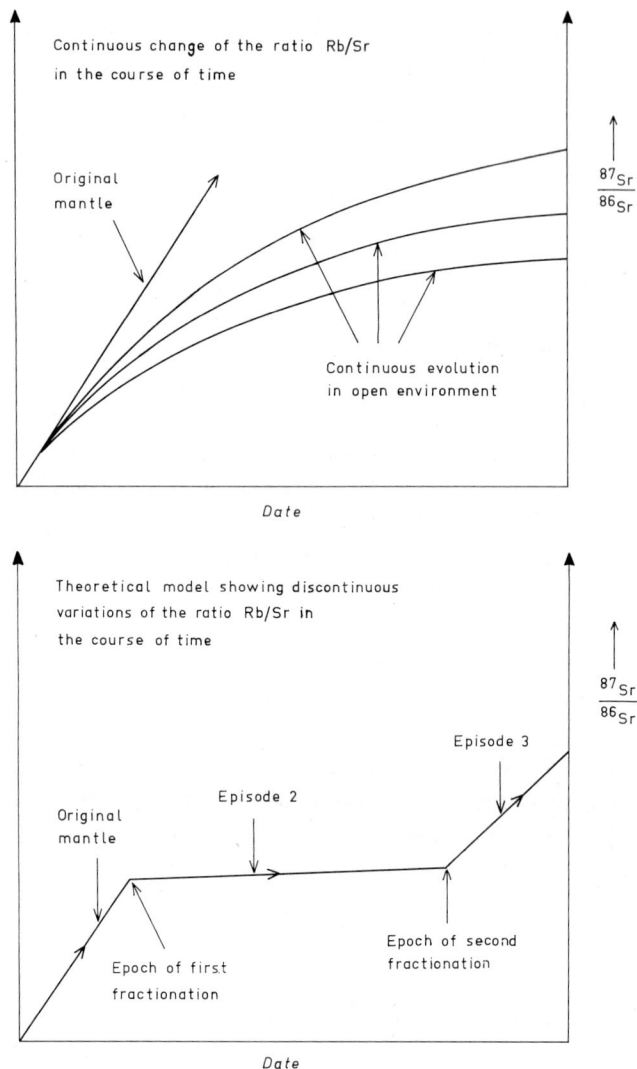

Fig. 7.14.   Schematic representation showing the possible evolution of the ratio $^{87}Sr/^{86}Sr$ in the mantle while the ratio of Rb/Sr varied.

All this is valid regardless of the relative richness in the mantle of U to Pb (richness defined by the ratio $^{238}U/^{204}Pb$). It is thus possible to test by this model the reality of the closed environment for the mantle. Gast (1958); Tilton and Grunfelder (1960); and Hedge; and later Tatsumoto (1966) have shown that alkaline basalts as well as tholeites do not give graph points on the line for the closed model. However, a large difference appears between these two types of basalts.

For tholeites, the ratios $^{206}Pb/^{204}Pb$ are fairly variable (although less than for alkaline basalts) but the ratios $^{207}Pb/^{204}Pb$ are practically constant.

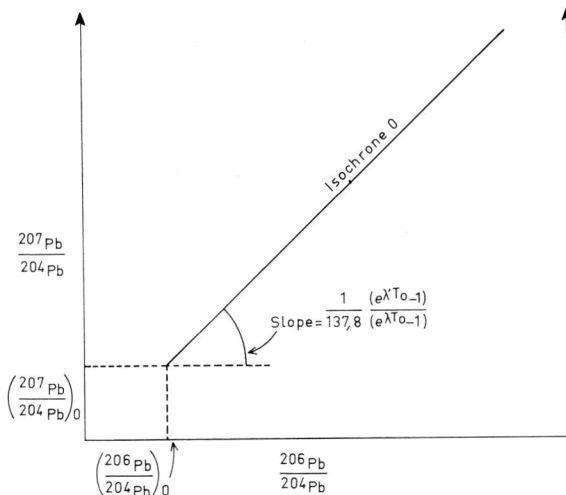

Fig. 7.15.   Principle of $^{207}Pb/^{204}Pb$ vs $^{206}Pb/^{204}Pb$ diagram.

For alkaline basalts the two isotopic ratios vary considerably (Figure 7-16).

Now it must be remembered that the radioactive constants of ^{238}U and ^{235}U are very different and that as a result, some differences in the ratios $^{207}Pb/^{204}Pb$ can only result from environmental differences that developed over 2.7×10^9 yr ago.

It thus seems at first glance that some ancient phenomena influenced the mantle from which alkaline basalts arose; oppositely, the domains where the tholeites arose were fractionated chemically more recently in geological history.

In fact, this interpretation is difficult to accept for numerous reasons, but the solution has not yet been found.

In any case, it can be seen that these studies yield irreplaceable information concerning the geochemical history of the mantle.

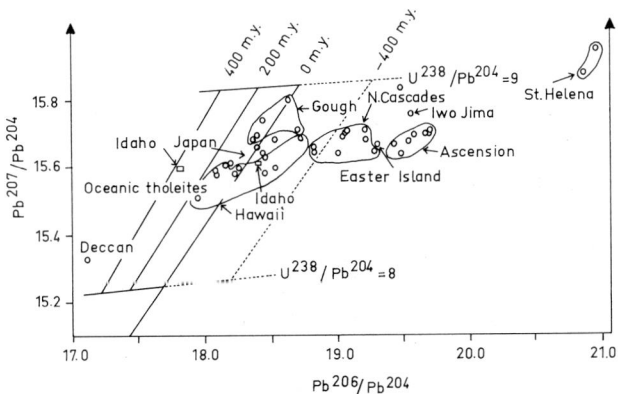

Fig. 7.16.   Position of subrecent basalts in the diagram of $^{207}Pb/^{204}Pb$ vs $^{206}Pb/^{204}Pb$ (after Gast and Tatsumoto works).

7.6. Evolution of the Initial Ratios of $(^{87}SR/^{86}Sr)$ and $(^{206}Pb/^{204}Pb)$ in Granites and the Formation of the Continents

It is known that geology and geophysics lead to a distinction between the continental gneiss-granite crust and the oceanic basalt crust. Now modern theories of sea floor spreading do not assign major importance to this distinction (the idea of lithospheric plates being more important than the notion of the crust) and they do not indicate at all how the continental crust may have formed. Examination of the isotopic composition of granites provides an opening to this problem. In order to discuss the data, let us establish several reference models.

It is imaginable at first that, after an initial period of formation, the crust then grew by a self-reproducing, conservative process. Erosion deposited sediments from the weathering of granites, and these sediments were metamorphosed during orogenic periods, then granitized, creating a new segment of continental crust.

This model can be formulated mathematically, for in this case, one may write as a first approximation:

$$(^{87}Sr/^{86}Sr)_t = (^{87}Sr/^{86}Sr)_i + (^{87}Rb/^{86}Sr)_{crust} (e^{\lambda_1 t_0} - e^{\lambda_1 t})$$
$$(^{206}Pb/^{204}Pb)_t = (^{206}Pb/^{204}Pb)_i + (^{238}U/^{204}Pb)_{crust} (e^{\lambda_2 t_0} - e^{\lambda_2 t}).$$

Opposing this model, a very different process might be imagined: with each orogenesis, a new piece of granitic crust formed by magmatic differentiation starting from the basic mantle. In this case one may write:

$$(^{87}Sr/^{86}Sr)_t = (^{87}Sr/^{86}Sr)_{t_0} + (^{87}Rb/^{86}Sr)_{mantle} (e^{\lambda_1 t_0} - e^{\lambda_1 t})$$
$$(^{206}Pb/^{204}Pb)_t = (^{206}Pb/^{204}Pb)_{t_0} + (^{238}U/^{204}Pb)_{mantle} (e^{\lambda_2 t_0} - e^{\lambda_2 t}).$$

This model is the model of continuous accretion from the mantle by the continents through geologic time.

Now the initial ratio $(^{87}Sr/^{86}Sr)_i$ can be measured for each granite as well as its age thanks to the use of the method of isotopic evolution lines applied to whole rock systems.

The initial ratio $^{206}Pb/^{204}Pb$ may be obtained in a similar fashion, but it is preferable to estimate it starting from potash feldspars that contain only very small amounts of U and are thus a good 'memento' of the initial isotopic composition of the lead.

Comparison of the results for North America shows a contradiction.

The data for the ratio $^{87}Sr/^{86}Sr$ favor a continuous growth of the continents (Hurley *et al.*, 1962) whereas those for the ratio $^{206}Pb/^{204}Pb$ favor a self reproduction since 3×10^9 yr ago (Patterson, 1964).

This contradiction can be resolved thanks to the following models.

Using the sea floor spreading hypothesis, Armstrong (1971) has proposed a model to resolve this paradox. The model involves a sinking sediment slab solidary of the lithosphere in a subduction zone. However, we know now that sediments are not buried down to the mantle (observation in the Indonesia trench) and also that sea floor spreading mechanism is probably limited back in time (Allègre, 1971). Allègre has proposed an alternative hypothesis in which mixing between mantle and crust occurs in geosyncline and also during the ascent of granitic precursors.

We suppose that all granite crust results from a mixing of a portion of mantle and an ancient, remobilized crust. The proportions of the mixture vary with time, and the mixing takes place largely in the orogenic trenches.

In this case, the Sr that is plentiful in the material of the mantle but by contrast is readily removed from granitic matter (erodable plagioclases) would lead to the expectation of augmented differentiation in the mantle. The Pb by contrast, less abundant in the basic matter and well conserved in the eroded granitic matter (potash feldspars and zircons) would tend to give weight to the idea of the remobilized crust.

Our model, while explaining this paradox, permits a quantitative resolution of the formation of the crust. The analysis shows in effect that the ratio of the mixture of crust-mantle grew from 0.2, 2.7×10^9 yr ago, to 10 today. It thus seems that the crust became formed from the mantle in early precambrian times and that actually, the continental granitic crust functions almost in a closed cycle.

References

Allègre, C. J.: 1967a, *Earth Planetary Sci. Letters* **2**, 57.

Allègre, C. J.: 1967b, *Introduction à la Géochronologie des Systèmes Ouverts*, Thesis, Paris.

Allègre, C. J.: 1971, in C. J. Allègre and Mattauer (ed.), *Structure et Dynamique de la Lithosphère*, Herman, Paris.

Damon, P. and Kulp, J. L.: 1958, *Am. Mineralogist* **43**, 433.

Hart, S. R.: 1964, *J. Geol.* **72**, 493.

Hurley, P. M. H., Hughes, G., Faure, H. W., Fairbain, W. H., and Pinson, W. H.: 1962, *J. Geophys. Res.* **67**, 5315.

Lanphere, M. A., Wasserburg, G. J., Albee, A. L., and Tilton, G. R.: 1964, *Isotopic and Cosmic Chemistry*, North-Holland, Amsterdam.

Nier, A. V.: 1939, *Phys. Rev.* **55**, 153.

Patterson, C. G.: 1964, *Isotopic and Cosmic Chemistry*, North-Holland, Amsterdam.

Silver, L. T.: 1963, *Radioactive Dating*, Vienna, IAEC.

Tatsumoto, M.: 1966, *Sciences* **153**, 1094.

Tilton, G. R.: 1960, *J. Geophys. Res.* **65**, 2933.

Wasserburg, G. J.: 1961, *Ann. N.Y. Acad. Sci.* **91**, 583.

Wasserburg, G. J.: 1963, *J. Geophys. Res.* **68**, 4823.

Wasserburg, G. J.: 1964, *Trans. Am. Geophys. Union* **45**, 407.

Wetherill, G. W.: 1956, *Trans. Am. Geophys. Union* **37**, 320.

INDEX

GEOPHYSICS AND ASTROPHYSICS MONOGRAPHS

AN INTERNATIONAL SERIES OF FUNDAMENTAL TEXTBOOKS